High-Temperature Superconductor Materials, Devices, and Applications

Technical Resources

Journal of the American Ceramic Society

www.ceramicjournal.org

With the highest impact factor of any ceramics-specific journal, the *Journal of the American Ceramic Society* is the world's leading source of published research in ceramics and related materials sciences.

Contents include ceramic processing science; electric and dielectic properties; mechanical, thermal and chemical properties; microstructure and phase equilibria; and much more.

Journal of the American Ceramic Society is abstracted/indexed in Chemical Abstracts, Ceramic Abstracts, Cambridge Scientific, ISI's Web of Science, Science Citation Index, Chemistry Citation Index, Materials Science Citation Index, Reaction Citation Index, Current Contents/ Physical, Chemical and Earth Sciences, Current Contents/Engineering, Computing and Technology, plus more.

View abstracts of all content from 1997 through the current issue at no charge at www.ceramicjournal.org. Subscribers receive full-text access to online content.

Published monthly in print and online. Annual subscription runs from January through December. ISSN 0002-7820

International Journal of Applied Ceramic Technology

www.ceramics.org/act

Launched in January 2004, *International Journal of Applied Ceramic Technology* is a must read for engineers, scientists,and companies using or exploring the use of engineered ceramics in product and commercial applications.

Led by an editorial board of experts from industry, government and universities, *International Journal of Applied Ceramic Technology* is a peer-reviewed publication that provides the latest information on fuel cells, nanotechnology, ceramic armor, thermal and environmental barrier coatings, functional materials, ceramic matrix composites, biomaterials, and other cutting-edge topics.

Go to www.ceramics.org/act to see the current issue's table of contents listing state-of-the-art coverage of important topics by internationally recognized leaders.

Published quarterly. Annual subscription runs from January through December. ISSN 1546-542X

American Ceramic Society Bulletin

www.ceramicbulletin.org

The *American Ceramic Society Bulletin*, is a must-read publication devoted to current and emerging developments in materials, manufacturing processes, instrumentation, equipment, and systems impacting the global ceramics and glass industries.

The *Bulletin* is written primarily for key specifiers of products and services: researchers, engineers, other technical personnel and corporate managers involved in the research, development and manufacture of ceramic and glass products. Membership in The American Ceramic Society includes a subscription to the *Bulletin*, including online access.

Published monthly in print and online, the December issue includes the annual *ceramicSOURCE* company directory and buyer's guide. ISSN 0002-7812

Ceramic Engineering and Science Proceedings (CESP)

www.ceramics.org/cesp

Practical and effective solutions for manufacturing and processing issues are offered by industry experts. CESP includes five issues per year: Glass Problems, Whitewares & Materials, Advanced Ceramics and Composites, Porcelain Enamel. Annual subscription runs from January to December. ISSN 0196-6219

ACerS-NIST Phase Equilibria Diagrams CD-ROM Database Version 3.0

www.ceramics.org/phasecd

The ACerS-NIST Phase Equilibria Diagrams CD-ROM Database Version 3.0 contains more than 19,000 diagrams previously published in 20 phase volumes produced as part of the ACerS-NIST Phase Equilibria Diagrams Program: Volumes I through XIII; Annuals 91, 92 and 93; High Tc Superconductors I & II; Zirconium & Zirconia Systems; and Electronic Ceramics I. The CD-ROM includes full commentaries and interactive capabilities.

High-Temperature Superconductor Materials, Devices, and Applications

Proceedings of the 106th Annual Meeting
of The American Ceramic Society,
Indianapolis, Indiana, USA (2004)

Editors
M. Parans Paranthaman
Paul N. Barnes
Bernhard Holzpfel
Yutaka Yamada
Kaname Matsumoto
John K.F. Yau

Published by
The American Ceramic Society
PO Box 6136
Westerville, Ohio 43086-6136
www.ceramics.org

ISBN 1-57498-181-1

Contents

YBCO Coated Conductors

Buffer Layers

Bulk Superconductors

Preface

Major advances have been made in the last seventeen years in high-temperature superconductor (HTS) research, resulting in increasing use of HTS materials in commercial and pre-commercial applications. These materials have in common the complexity of their multi-component chemistry. Consequently it is not surprising that many aspects of the interplay between microscopic structure, macroscopic properties, and processing are still not fully understood. This symposium investigated the relationship between features at atomic level such as oxygen vacancies, stacking faults and site order/disorder, grain boundaries, film-substrate interactions, buffer-superconductor interactions, thermodynamic, transport, and other macroscopic properties. Phase diagrams, which are based on the relevant crystal chemistry, serve as a framework for interpreting microscopic/macroscopic interactions. The focus of this symposium was on the integration of crystal chemistry, phase equilibria and thermodynamics of high T_c materials in bulk, thin film and single crystal forms. This symposium also focused on key materials and processing developments in both MgB_2 and YBCO conductors that promise to broaden the commercial applications of HTS wire and thin film technology. The development of second-generation YBCO coated conductors continues to show the steady improvement toward the long-length processing capabilities. This symposium also covered fundamental material properties studies, new growth methods, device and materials integration research, and developments in designing and growing new materials, all involving epitaxial superconducting thin films.

This volume contains proceedings of the papers presented at the High-Temperature Superconductor Materials, Devices and Applications Symposium during the 106th Annual Meeting of the American Ceramic Society (ACerS), April 18-21, 2004 in Indianapolis, Indiana. This symposium was well supported with 42 papers presented. One poster session was well attended and provided a dynamic forum for discussions of many of the issues raised during the meeting. This symposium provided overviews of the Long Length Processing of YBCO Coated Conductors, Bulk Processing and Flux Pinning, In-situ Diagnostics and Reel to-Reel Characterizations, Crystal Chemistry and Non-YBCO Materials, Biaxially Textured Templates (RABiTS, IBAD, and ISD), New Developments in Buffer Layer Technology. This volume is divided into three sections: (1) YBCO coated conductors, (2) Buffer layers, and (3) Bulk superconductors. The order in which the papers appear in this volume and the division into which they are organized may be different from that of the presentations at the meeting. It is hoped that this comprehensive volume will be a good summary of the latest developments in high-temperature superconductor research.

We acknowledge the service provided by the session chairs and appreciate the valuable assistance from the ACerS programming coordinators. We are also in debt to Greg Geiger for his involvement in editing and producing this book. Special thanks are due to the speakers, authors, manuscript reviewers, and ACerS officials for their contributions. The financial support of this meeting was provided by Electronics Division.

M. Parans Paranthaman
Paul N. Barnes
Bernhard Holzpfel
Yutaka Yamada
Kaname Matsumoto
John K.F. Yau

YBCO Coated Conductors

IMPROVING FLUX PINNING IN YBa$_2$Cu$_3$O$_7$ COATED CONDUCTORS BY CHANGING THE BUFFER LAYER DEPOSITION CONDITIONS

B. Maiorov, H. Wang, P. N. Arendt, S. R. Foltyn and L. Civale
Superconductivity Technology Center, MS K763,
Los Alamos National Laboratory, Los Alamos, NM 87545, USA

ABSTRACT

We present a comparative study of the flux pinning properties of YBa$_2$Cu$_3$O$_7$ films deposited by pulsed laser deposition on polycrystalline metal substrates with a biaxially oriented MgO template produced by ion-beam-assisted deposition (IBAD), varying the deposition temperature (T_{STO}) for the SrTiO$_3$ buffer layer. We find that when $T_{STO} = T^*_{STO} = 820°C$, the critical current density at *self-field* (J_c^{sf}) is maximized and the surface roughness minimized. On the contrary, *in-field* critical current density (J_c) measurements show that at high fields, samples with $T_{STO} < T^*_{STO}$ present higher J_c. Angular-dependent J_c measurements show that this improvement is due to the presence of additional extended parallel defects (in particular, thread dislocations), which produce a peak in J_c centered in the c-axis direction. We use the field dependence of the height and the width of this peak to compare the density of the correlated defects in samples prepared with different T_{STO}.

INTRODUCTION

Achieving high critical currents in YBa$_2$Cu$_3$O$_7$ (YBCO) films on metallic polycrystalline substrates is an essential step to make coated conductors (CC) a technological reality, being the ultimate goal for CC to obtain the highest possible critical current (I$_c$) in Amperes for a given tape width. There are two ways to increase I$_c$. The first (and one would think simplest) way is to grow thicker films. The second one is to increase J_c for a given thickness. The first approach carries material problems such as porosity[1], particulates and secondary phases, and other more intrinsic problems such as the thickness dependence of the critical current density J_c.[2] So, increasing the thickness alone is not a viable way, therefore it is extremely important to find out how to improve pinning in thick films. Historically, the highest J_c in YBCO have been obtained for thin films on single crystal substrates (SCS), with values in the MA range at liquid N$_2$ temperature. This is mainly due to the large density of defects in the films, which act as strong vortex pinning centers. However, for CC, it was always found that the J_c was lower than what could be obtained for a film of the same thickness deposited on SCS.[2] The inferior performance of the CC was due to the diminished J_c at the low angle grain boundaries of the material[3,4] as compared to the intra-grain J_c, a fact that at least qualitatively has been well understood since the pioneer work on YBCO films on bi-crystals.[5] Thus, for several years the goal for CC was to obtain J_c as large as for films on SCS.

This milestone was accomplished by Foltyn *et al.*[6], who showed that CC made by pulsed laser deposition (PLD) on polycrystalline Hastelloy™, using an MgO template grown by ion beam assisted deposition (IBAD), have J_c at 75.5 K and *self-field* as large as those of films of the same thickness on SCS.[6] The main reason for the improvement is the better texture of the IBAD template, which leads to an in-plane texture of the YBCO better than 3° (defined by the full width at half maximum [FWHM] of the X-ray Φ scan peak), low enough to eliminate the detrimental effect of the higher-angle grain boundaries on J_c. An important implication of that study is that the current carrying capability of those highly textured CC is limited by intra-grain vortex pinning, allowing the use of simple macroscopic transport measurements to compare and contrast the temperature (T) and magnetic field (H) dependence of their J_c with that of films on SCS, to investigate what type of defects are controlling pinning in each case.

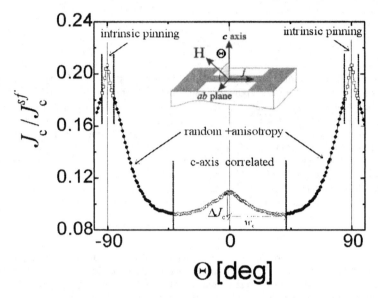

Figure 1: Angular pinning regimes for an YBCO on STO sample at $\mu_0 H = 1$ T and 75.5 K. The three regimes are: c-axis correlated pinning ($\Theta \sim 0°$) open circles, random + anisotropy regime ($40° < |\Theta| < 85°$) full circles, intrinsic pinning ($|\Theta| \sim 90°$) open squares. Insert: Sample geometry and magnetic field configuration.

We performed such a study in a previous paper.[7] We obtained particularly valuable information from the dependence of J_c on the orientation of H, which allowed us to discern between uncorrelated or correlated disorder and, in the latter case, to determine the orientation of the extended pinning structures. At 75 K, we were able to identify three angular regimes where pinning is dominated by different mechanisms, as depicted in Fig. 1. First there is a contribution coming from random point-like defects. Due to the mass anisotropy of YBCO, this defects produce a J_c that has a maximum at the ab-planes ($|\Theta| = 90°$), and a minimum at the c-axis ($\Theta = 0°$). These defects dominate in the range $40° < |\Theta| < 85°$ for this field. The presence of the ab-planes themselves also induces a sharp extra peak at $|\Theta| \sim 90°$, which is better observed when H is increased or T decreased. Thirdly, we find a peak centered at the c-axis due to correlated defects, such as dislocations and twin boundaries ($|\Theta| < 40°$). The three regimes are present in PLD YBCO films deposited on both SCS and high quality IBAD substrates. Over large field and angular ranges, J_c is higher in the coated conductors, demonstrating that the J_c of the films on single crystal substrates is not an upper limit for their performance.[7]

Table I: Data for the samples, sample number (#), YBCO thickness (t), deposition temperature of the STO buffer layer (T_{STO}), critical temperature (T_c) and *self-field* critical current density J_c^{sf}.

#	t [μm]	T_{STO} [°C]	$\Delta\Phi$ [deg]	T_c [K]	J_c^{sf} [MAcm^{-2}]
136	4.1	820	1.87	92.0	1.33
132	5.0	760	2.96	91.0	1.21
100	4.3	700	2.70	89.0	0.98

These findings provide a framework in which now we can compare films grown in different conditions, to try to correlate the microstructure with the pinning properties. Of course the main goal is to improve pinning, and in that regard many approaches are potentially interesting. For instance, a simple method of Rare Earth substitution changing the variance of the ionic radius[8] improves J_c at low fields by introduction of random defects. Even more attractive is the possibility to introduce additional c-axis correlated defects, particularly thread dislocations, which are known to be strong pinning centers.[9] In pure YBCO films the density of dislocations is dominated by the growth island size and their spacing is estimated to be rather large (~100 nm-500 nm).[9,10,11] That density can be increased by artificially induced precipitates through laser deposition of nanocrystallites of Y_2O_3,[10] which act as nucleation centers for additional dislocations, but this requires one extra step during the deposition. That disadvantage can be avoided by the inclusion of nanoparticles of $BaZrO_3$ in the target, which leads to superconducting films with significantly increased in-field J_c due to the extra dislocations induced by the presence of these particles in the YBCO matrix.[15]

It would be desirable to increase the density of pinning centers in CC made with plain YBCO, by far the most studied and well characterized material. One obvious way to increase the dislocation density would be to decrease the island size, which bounds the dislocations,[11] e.g. by reducing growth temperature, but this is non-trivial because the crystalline quality of the film would be compromised. As a more indirect but potentially simpler alternative, we can explore the possibility to modify the growing conditions of the layers beneath the YBCO in the CC (chemical barrier layer, IBAD texturing and buffer layer), to try to improve the pinning structures in the superconducting film. The natural first choice is the buffer layer in immediate contact with the YBCO.

The objective of this paper is to investigate the effects of the $SrTiO_3$ (STO) buffer layer on the pinning properties of YBCO. The first part deals with the effects on the structural properties and *self-fields* J_c. The second part focuses on the *in-field* J_c and its angular behavior, making emphasis on the very thick samples (~5 μm).

EXPERIMENTAL

To ensure that no substrate effects were introduced, all the substrates used in this experiment were cut from the same meter-long Hastelloy™ C276 tape with continuously processed MgO templates that consist primarily of an IBAD-MgO layer and a homo-epitaxial MgO layer. Detailed MgO template processing conditions can be found elsewhere.[12,13]

The depositions of the 60 nm thick STO buffer layer and YBCO films (from 1 to 5 μm thick) were performed by pulsed laser deposition (PLD) with a KrF excimer laser ($\lambda = 248$ nm).

The substrate's temperature for STO growth (T_{STO}) was varied from 670°C to 860°C at constant optimized oxygen pressure.[14] YBCO films were deposited thereafter on these buffers at an optimized substrate temperature of 760°C and O_2 pressure of 200 mTorr. After deposition samples were cooled to room temperature in O_2 at 300 Torr.

X-ray diffraction analysis including normal θ~2θ scan, Φ scan and rocking curve, were performed. Scanning electron microscopy (SEM) and atomic force microscopy (AFM) were used to study the surface morphology and roughness of as-grown YBCO and STO buffer layers for various deposition conditions. Cross-sectional microstructure studies of these as-deposited films, including selected area diffraction (SAD), transmission electron microscopy (TEM) and high resolution TEM, were performed with a JEOL-3000F analytical electron microscope with point-to-point resolution of 0.17 nm.

The critical temperature (T_c) was measured inductively. Transport J_c measurements were performed on bridges using a four-probe technique and a 1 μV/cm voltage criterion, with the films immersed in liquid N_2 (75.5 K). See Table I for sample parameters. The samples were rotated in a way that H was always perpendicular to the current (see insert Fig. 1), in order to assure a maximum Lorentz force configuration ($J \perp H$). The angle Θ between H and the normal to the films (which coincides with the crystallographic c axis) was determined to better than 0.1°.

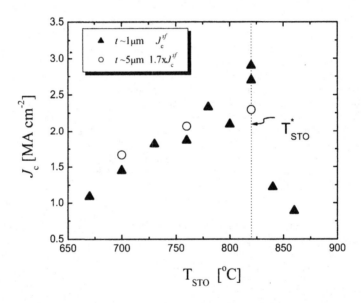

Figure 2: Critical density current measured at *self-field* (J_c^{sf}) as a function of the temperature of deposition of the STO buffer layer (T_{STO}) for 1.3 μm thick samples (full triangles) and the samples in Table I (open circles multiplied by 1.7 factor). The dotted line marks $T^*_{STO} = 820°C$, where J_c^{sf} is maximized.

RESULTS AND DISCUSSION

Self-Field and Microscopy Results

In Fig. 2 we plot J_c^{sf} as a function of T_{STO}. As T_{STO} increases from 670°C to 820°C, J_c^{sf} increases continuously from about 1 MAcm^{-2} to 3 MAcm^{-2}. If T_{STO} is further increased to 860°C, J_c drops to 0.9MA cm^{-2}. Thus, there is an optimum temperature (in terms of J_c^{sf}) for growing the STO buffer layer, $T^*_{STO} \sim 820$°C. This is an important result. The in-plane texture, obtained from an X-ray Φ-scan on the YBCO (103) peak, was measured and found to be almost constant with a value close to 3°. This indicates that J_c variations in this films do not arise from the YBCO texture, and other parameter(s) must be playing an important part in determining J_c^{sf}.

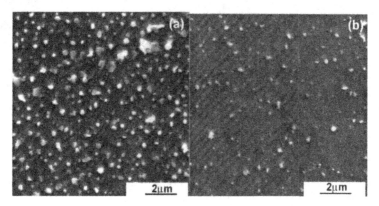

Figure 3: SEM image of the STO buffer surface for different deposition temperatures a)T_{STO} = 670°C b) T_{STO} = 820°C

In order to understand how the STO deposition temperature has such a strong effect on J_c, surface morphology and roughness studies of STO and YBCO films were performed by SEM and AFM. Figures 3 (a) and (b) show and contrast the surface morphology of STO buffer layers deposited at 670°C and 820°C respectively. At 820°C, the surface of the STO buffer layer is much smoother, with fewer particles. On the surface corresponding to T_{STO}=670°C, a higher density and bigger size of the particles is found. These particles are made of STO and appear white due to their height. The particle size ranges from 200-400 nm in this image. Analogous images for the buffer layers deposited at the other temperatures indicate that the surface of STO grown at 820°C is cleaner and smoother than the ones deposited at lower T_{STO}. This observation about the roughness is confirmed by AFM measurements. Over an area of 10μm×10μm the average surface roughness of STO buffer layers deposited at 670°C, 760°C and 820°C is about 12 nm, 6nm and 4 nm, respectively.[14] As T_{STO} is further increased to 860°C, not only does the STO surface get rougher, but also grain boundaries of the metal substrate begin to appear. This suggests that, at such high temperatures, grain-boundary diffusion from the metal substrate affects the growth of the STO buffer layer.

We have also performed the same experiment on STO-buffered single crystal MgO substrates, with similar results. From these observations, it is clear that STO growth conditions have strong effects on the superconducting properties of YBCO films, but not on the texture of YBCO. For more details on the structural studies, please see Ref. 14.

Now we turn to study the nature of these particles. In Fig. 4, we present TEM micrographs of the STO/YBCO interfaces for buffer layers grown at 820°C and 670°C. For T_{STO} = 820°C, the clean interface between YBCO and STO is seen, while for T_{STO} = 670°C a transversal view of one of the particles is shown. It is interesting to note that these small STO particles are crystalline and follow the orientation of the STO buffer, and that the overall in-plane

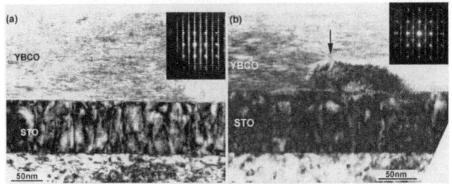

Figure 4: Cross-sectional low magnification bright field TEM images of YBCO on MgO(100) substrate with STO buffer layer deposited at (a) 820°C and (b) 670°C. Corresponding SAD patterns are shown as insets in (a) and (b). Arrow marked in (b) corresponds to a low angle boundary in YBCO above an STO outgrowth

alignment of YBCO is not affected by these epitaxial STO outgrowths. Besides the small particles, when T_{STO} is decreased, low angle grain boundaries (less than 1°) can be found in the YBCO next to the particles, as seen in Fig. 4b, but previous evidence strongly suggests[3,4,5] that such low angle grain boundaries should not significantly depress J_c. So, in summary, the correlation between the proliferation of the STO outgrowths and the decrease in J_c^{sf} is clear, but the reason for that correlation is unknown.

In-Field Results

Although in terms of J_c^{sf} the STO particles on the buffer layer are detrimental, one may wonder whether they may generate additional pinning and thus have a positive effect on the field dependence of J_c. Specifically, we expect these particles to enhance the density of dislocations, in analogy with the already mentioned surface distribution of Y_2O_3 nanoparticles. The presence of $BaZrO_3$ nanoparticles also induces a proliferation of thread dislocations in the YBCO films, which results in a large enhancement in the c-axis peak of J_c.[15]

Since one of our objectives is to improve the pinning to allow higher values of total critical current, we decided to start our study of the influence of the STO particles on the *in-field* J_c on the very thick films (~ 5 µm), listed in Table I. The study of the thinner samples is currently underway.

Before we start the in-field study, in Fig. 2 we include the J_c^{sf} as a function of T_{STO} for three thick films grown on the same batch of IBAD, shown in open symbols. In order to facilitate the comparison of all the data in the same plot, the J_c^{sf} values of the 5 µm thick films are multiplied by 1.7, to approximately compensate for the thickness dependence. For these thicker films we observe the same trend in J_c^{sf} as a function of T_{STO}, with a maximum at $T_{STO}^* \sim 820°C$, even though the variations are not as big as in the thinner samples of Fig. 2, being J_c^{sf} of #100 ($T_{STO} = 700°C$) approximately 70% of #136 ($T_{STO} = 820°C$). It is worth noticing that these thick samples carry the equivalent of more than 600 A/cm at 75.5K and self field.

In Fig. 5a we present the field dependence of J_c for samples #100 and #136. We observe that, although sample #100 has a lower J_c^{sf}, for $H//c$ it exhibits a slower decay of J_c as a function

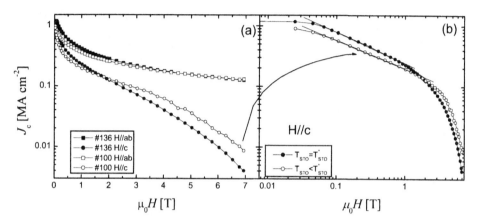

Figure 5: a) Field dependence of samples #100(open symbols) and #136(full symbols) for $H///ab$ (squares)and $H//c$ (circles) at T = 75.5 K.

of H. As a result, around 2 T the J_c values for both films are equal, and for higher fields J_c(#100) $> J_c$(#136), this difference being at least a factor of 2 at 7 T. The relative improvement is less significant for $H//ab$, where both samples have almost identical J_c at high fields, indicating that the primary effect of the STO particles is to enhance J_c in the c-axis direction. This is confirmed by the qualitatively different behavior of the ratios J_c(#100) / J_c(#136) for $H//c$ and $H//ab$, as shown in Fig. 6 a and b respectively.

If we plot $J_c(H)$ curves for $H//c$ in a log-log scale (Fig. 5b), in both curves we observe the usual $J_c \sim H^\alpha$ dependence in the low field region. For samples #132 (not shown) and #136, α presents a typical value of 0.55,[7,9,16] while for sample #100 we find $\alpha=0.46$. That is, the improved pinning associated with the STO particles manifests as a reduced α. The same trend has been found when adding $BaZrO_3$ particles.[15] It is natural to expect a smaller α upon increasing the dislocation density and keeping the rest of the parameters fixed, since the vortices have more available pinning centers. Previously, Klaasen et al.[16] found the opposite effect, i.e., α increased upon increase in the number of dislocations. In that work, however, in order to change the number the dislocations they changed the deposition conditions of the YBCO, which were kept constant in our experiments.[14,15]

Definite evidence that the pinning increase for H//c is due to correlated defects along the c axis comes from the angular dependence of J_c, plotted for different fields in Fig. 7. At 1 T (Fig. 7a) one can see that #136, grown at T^*_{STO}, has higher J_c, throughout all the angular range. However, at 3 T the increase of J_c near the c axis for #100 is evident, while for larger $|\Theta|$ the two curves have similar values and angular dependences. In a previous paper[7] we have shown that in PLD samples, for $\mu_0 H > 1$ T the height of the c-axis peak (ΔJ_c, see definition in Fig. 1) decays with an exponential law as

$$\Delta J_c = J^* \exp(- H/H_0). \qquad (1)$$

where the constant J^* was sample dependent, but the parameter H_0 was almost sample independent in all the PLD YBCO films on either IBAD MgO or SCS investigated, with $\mu_0 H_0 = 1.6 \pm 0.2$ T. When this analysis is done in the present samples (see Figure 8.b), we find $\mu_0 H_0 = 1.5$ and 1.55 T for #136 and #132 respectively, while sample #100 has a $\mu_0 H_0 = 2.1$ T, which is significantly higher and out of the error bars.

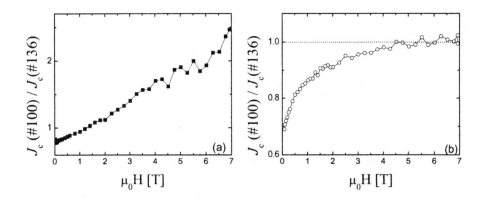

Figure 6: Ratio of J_c of sample #100 and #136 at 75.5K for the main crystallographics directions a) H//c full squares b) H//ab open circles.

We also found a systematic enhancement of the c-axis peak width w_c (see definition in Fig. 1) as in Fig. 8a, which indicates that the angular range of influence of the correlated defects has been extended. It is worth noticing that the definition of ΔJ_c and w_c are a conservative estimation of the height and width of the peak due to correlated defects. Indeed, in Fig. 7b, one can see that film # 100 has a J_c higher than film # 136 up to $\Theta \sim 50°$, indicating that the correlated defects in film # 100 act until this angle, and not up to the minimum which defines w_c (32° and 25° for #100 and #136 respectively). Nevertheless, w_c is a useful parameter, specially for comparison purposes.

In summary, the dislocations associated with the STO particles improve pinning both in the low field ($\mu_0 H < 1$ T) and high field ($\mu_0 H > 1$ T) regimes, as manifested by the decrease of α and the increase of H_0 and w_c, respectively. These results are similar to those found in YBCO films with added BaZrO$_3$ nanoparticles.[9] We note that the size, density, and spatial distribution of both types of particles are very different. In order to optimize the pinning effects of the resulting dislocations, a more quantitative study of the microstructure in each case is necessary.

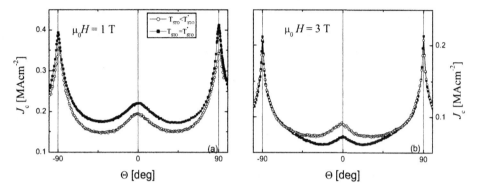

Figure 7: Angular dependence of J_c for sample #100 (T$_{STO}$=700°C open circles) and #136 (T$_{STO}$= T*$_{STO}$ = 820°C full squares) at a constant magnetic field intensity of a) $\mu_0 H = 1$ T b) $\mu_0 H = 3$ T.

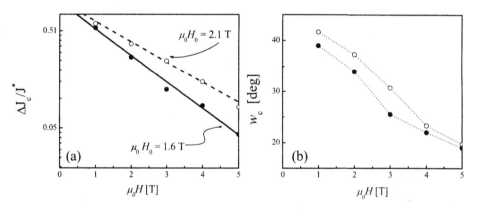

Figure 8: normalized ΔJ_c and w_c as a function of H in a) and b) respectively. For sample #100 (open circles) and #136 (full circles).

CONCLUSIONS

We have grown high quality superconducting YBCO thick films on polycrystalline metal substrates with IBAD MgO using a thin STO buffer layer. The pulsed laser deposition growth conditions of the buffer layer have a strong effect on the superconducting properties of the YBCO. In particular, for a certain buffer layer deposition temperature T^*_{STO}, the J_c^{sf} is maximized, and the surface roughness is minimized. However, a lower T_{STO} results in an improvement of the in-field J_c due to the appearance of additional c-axis oriented dislocations, which are associated with the proliferation of epitaxial STO particles at the YBCO/STO interface.

This is a very simple and inexpensive way to increase the in-field pinning in coated conductors, without incorporating an additional processing step, and even maintaining the YBCO deposition conditions unchanged. Our results demonstrate that, once the detrimental effects of the grain boundaries have been eliminated, it is possible to further improve the already high pinning in YBCO films. In principle, this method is independent of the superconducting film to be deposited. Thus, it could be applied to other RE-123 films, as well as in combination with rare earth substitutions, nanoparticle inclusions, or other methods to introduce pinning centers.

ACKNOWLEDGEMENTS

This work is supported by U.S. Department of Energy.

[1] Q. X. Jia, S. R. Foltyn, P. N. Arendt, and J. F. Smith "High-temperature superconducting thick films with enhanced supercurrent carrying capability" Appl. Phys. Lett. **80**, 1601 (2002).

[2] S. R. Foltyn, P. Tiwari, R. C. Dye, M. Q. Le, and X. D. Wu, "Pulsed laser deposition of thick $YBa_2Cu_3O_{7-\delta}$ films with $J_c \gtrsim 1$ MA/cm^2" Appl. Phys. Lett. **63**, 1848 (1993).

[3] D. T. Verebelyi, D. K. Christen, R. Feenstra, C. Cantoni, A. Goyal, D. F. Lee, M. Paranthaman, P. N. Arendt, R. F. DePaula, J. R. Groves, and C. Prouteau, "Low angle grain boundary transport in $YBa_2Cu_3O_{7-\delta}$ coated conductors" Appl. Phys. Lett. **76**, 1755 (2000); D. M. Feldmann, D. C. Larbalestier, D. T. Verebelyi, W. Zhang, Q. Li, G. N. Riley, R. Feenstra, A. Goyal, D. F. Lee, M. Paranthaman, D. M. Kroeger, and D. K. Christen, "Inter- and intragrain transport measurements in $YBa_2Cu_3O_{7-x}$ deformation textured coated conductors" *ibid* **79**, 3998 (2001).

[4] D. T. Verebelyi, C. Cantoni, J. D. Budai, D. K. Christen, H. J. Kim, and J. R. Thompson, "Critical current density of $YBa_2Cu_3O_{7-\delta}$ low-angle grain boundaries in self-field" Appl. Phys. Lett. **78**, 2031 (2001).

[5] P. Chaudhari, J. Mannhart, D. Dimos, C. Tsuei, J. Chi, M. Oprysko, and M. Scheuermann, "Direct measurement of the superconducting properties of single grain boundaries in $Y_1Ba_2Cu_3O_{7-\delta}$", Phys. Rev. Lett. **60**, 1653 (1988); D. Dimos, P. Chaudhari, J. Mannhart, and F. K. Legoues, "Orientation Dependence of Grain-Boundary Critical Currents in $YBa_2Cu_3O_{7-\delta}$ Bicrystals" *ibid*. **61**, 219 (1988).

[6] S. R. Foltyn, P. N. Arendt, Q. X. Jia, H. Wang, J. L. MacManus-Driscoll, S. Kreiskott, R. S. DePaula, L. Stan, G. R. Groves, and P. C. Dowden, "Strongly coupled critical current density values achieved in $Y_1Ba_2Cu_3O_{7-\delta}$ coated conductors with near-single-crystal texture" Appl. Phys. Lett. **82**, 4519 (2003).

[7] L. Civale, B. Maiorov, A. Serquis, J. O. Willis, J. Y. Coulter, H. Wang, Q.X. Jia, P.N. Arendt, J.L. MacManus-Driscoll, M.P. Maley, and S.R. Foltyn, "Angular-dependent vortex pinning mechanisms in $YBa_2Cu_3O_7$ coated conductors and thin films" Appl. Phys. Lett. **84**, 2121 (2004).

[8] J. L. MacManus-Driscoll, S. R. Foltyn, Q. X. Jia, H. Wang, A. Serquis, B. Maiorov, L. Civale, Y. Lin, M.E. Hawley, M.P. Maley and D. E. Peterson "Systematic Enhancement of in Field-Critical Current Density with Rare Rarth Ion Size Variance in Superconducting Rare Earth Barium Cuprate Films" to be published in Appl. Phys. Lett.

[9] B. Dam , J. M. Huijbregtse, F. C. Klaassen, R.C.F van der Geest, G. Doornbos, J.H. Rector, A.M. Testa, S. Freisem, J.C. Martinez,B. Stauble-Pumpin and R. Griessen, "Origin of high critical currents in $YBa_2Cu_3O_{7-\delta}$ superconducting thin films", Nature **399** (675) 439-442 (1999)

[10] J.M. Huijbregtse,B. Dam, R.C.F. van der Geest, F.C. Klaassen, R. Elberse, J.H. Rector, and R. Griessen, "Natural strong pinning sites in laser-ablated $YBa_2Cu_3O_{7-\delta}$ thin films " Phys. Rev. B. **62**, 1338 (2000).

[11] B. Dam, J.M. Huijbregtse and J.H. Rector, "Strong pinning linear defects formed at the coherent growth transition of pulsed-laser-deposited $YBa_2Cu_3O_{7-\delta}$ films" Phys. Rev. B **65**, 064528 (2002).

[12] J. R. Groves, P. N. Arendt, S. R. Foltyn, Q. X. Jia, T. G. Holesinger, H. Kung, E. J. Peterson, R. F. DePaula, P. C. Dowden, L. Stan, and L. A. Emmert, "High critical current density $YBa_2Cu_3O_{7-\delta}$ thick films using ion beam assisted deposition MgO bi-axially oriented template layers on nickel-based superalloy substrates", J. Mater. Res. **16**, 2175 (2001).

[13] J. R. Groves, P. N. Arendt, S. R. Foltyn, Q. X. Jia, T. G. Holesinger, H. Kung, R. F. DePaula, P. C. Dowden, E. J. Peterson, L. Stan, and L. A. Emmert, "Recent progress in continuously processed IBAD MgO template meters for HTS applications" Physica C **382**, 43 (2002).

[14] H. . Wang, S. R. Foltyn, P. N. Arendt, Q. X. Jia, J. L. MacManus-Driscoll, L. Stan, Y. Li, X. Zhang and P. C. Dowden "Microstructure of $SrTiO_3$ buffer layers and its effects on superconducting properties of $YBa_2Cu_3O_{7-\delta}$ coated conductors" Submitted to J. Mater. Res.

[15] J. L. MacManus-Driscoll, S. R. Foltyn, Q. X. Jia, H. Wang, A. Serquis, B. Maiorov, L. Civale, Y. Lin, M.E. Hawley, M.P. Maley and D. E. Peterson "Strongly Enhanced Current Densities in Superconducting Coated Conductors of BaZrO3-Doped $YBa_2Cu_3O_{7-x}$ " Submitted to Nature Materials

[16] F. C. Klaasen, G. Doornbos, J. M. Huijbregtse, R. C. F. van der Geest, B. Dam, and R. Griessen, "Vortex pinning by natural linear defects in thin films of $YBa_2Cu_3O_{7-\delta}$" Phys. Rev. B **64**, 184523 (2001).

PROCESSING AND CHARACTERIZATION OF $(Y_{1-x}Tb_x)Ba_2Cu_3O_{7-z}$ SUPERCONDUCTING THIN FILMS PREPARED BY PULSED LASER DEPOSITION

J. W. Kell, T. J. Haugan, P. N. Barnes, M. F. Locke, and T. A. Campbell
Air Force Research Laboratory
Propulsion Directorate
2645 Fifth St., Ste. 13 Bldg. 450
Wright-Patterson AFB, OH 45433-7919, USA

C.V. Varanasi and L. B. Brunke
University of Dayton Research Institute
Air Force Research Laboratory
2645 Fifth St., Ste. 13 Bldg. 450
Wright-Patterson AFB, OH 45433-7919, USA

ABSTRACT

$REBa_2Cu_3O_{7-z}$ (RE123) superconductors are being considered for applications of thin film coated conductors because of their high critical transition temperatures (T_c) (~ 92 K), and high critical current density (J_c) at 77 K in applied magnetic fields. This paper considers the partial substitution of Tb in the $(Y_{1-x}Tb_x)Ba_2Cu_3O_{7-z}$ ((Y,Tb)123) structure (x = 0.01 and 0.1) to enhance flux pinning. Tb has normal valence states of +3, +4 which can substitute for Y which has a normal valence state of +3. The crystal ionic radii of $Tb^{(+3)}$ is 1.04 Å which is quite close to the ionic radii of $Y^{(+3)}$ of 1.02 Å. Tb123 normally is processed as a non-superconducting phase; therefore, substitution of Tb123 for Y123 into a bulk or thin film superconductor has the potential to create localized regions of size on the order of one unit cell or larger of either reduced T_c regions or potential site defects. Such regions can provide the non-superconducting pinning centers with particle densities approaching $1.5\text{-}3 \times 10^{11}$ cm^{-2} which are necessary to pin magnetic fields of ~3-6 T. This paper considers the properties of (Y,Tb)123 thin films prepared by pulsed laser deposition. Properties of the films including T_c and the magnetic field dependence of J_c at 77 K will be presented.

INTRODUCTION

Coated conductor technology for biaxilly aligned $YBa_2Cu_3O_{7-\delta}$ (Y123) on buffered metallic substrates with $J_c > 1$ MA/cm^2 offer great promise as a second generation high temperature superconducting (HTS) wire for use in generators and motors [1-8]. Y123 has many useful properties at 77 K such as high critical current densities (J_c) and good flux pinning in applied magnetic fields, which is critical in most applications [1-3]. When magnetic fields are applied parallel to the c-axis, J_c will typically decrease by a factor of 10 to 100 within the range of 1 T < B_{appl} < 5 T [9,10]. As such, further improvement of J_c is important, especially for c-axis orientation of the applied field, to allow further reduction in system weight and size. In most applications, the value, J_c(H) places an upper limit on the magnetic field that can be produced/applied for a given coil design.

Since Y123 is a type II superconductor, the best way to increase flux pinning is to introduce a high density of non-superconducting defects into the material [1-4]. One way to accomplish this is to introduce a second phase into the Y123 that is not superconducting. In this work,

$TbBa_2Cu_3O_7$ (Tb123) was incorporated directly into the 123 matrix to increase the flux pinning. Tb123 was considered as a pinning agent for several reasons. First, terbium does not degrade the T_c of Y123 [11]. Also, the divalent nature of terbium (+3 and +4 valence states) may allow for Tb^{4+} to act as pinning sites by alternate chemical bonding. Also, assuming a fully homogenous structure, the size of the pinning particles would effectively be the size of the Tb123 unit cell. It was also believed that small Tb doping of Y123 would allow for the same deposition parameters as in high quality Y123 during PLD. This would allow for the same O_2 pressures and laser fluences to be used, thereby reducing the processing time and simplifying the depositions. Previous studies have focused on substituting large quantities of Tb for Y in Y123 ($x \geq 0.1$) [11,12], whereas this study focuses on substituting small quantities ($x \leq 0.1$) of Tb for Y in Y123 thin films to achieve the appropriate defect densities if the Tb-inclusions are sufficiently dispersed.

EXPERIMENTAL

The laser ablation targets were manufactured in-house. Powder of a composition of $Y_{0.9}Tb_{0.1}Ba_2Cu_3O_{7-x}$ was prepared from Y_2O_3, $BaCO_3$, CuO, and Tb_4O_7 powder (all nominally 99.99+% pure). The powders were dried, mixed, and then calcined at 850°C and 880°C. This powder was then used to make two targets of compositions $Y_{0.9}Tb_{0.1}Ba_2Cu_3O_{7-x}$ and $Y_{0.99}Tb_{0.01}Ba_2Cu_3O_{7-x}$, with the second composition consisting of 10 mol % $Y_{0.9}Tb_{0.1}Ba_2Cu_3O_{7-x}$ and 90 mol % commercial Y123 powder. The targets were then fully reacted at 940°C and 920°C respectively for 50 hours. The $Y_{0.9}Tb_{0.1}Ba_2Cu_3O_{7-x}$ target was found to be 84.8% dense and the $Y_{0.99}Tb_{0.01}Ba_2Cu_3O_{7-x}$ target was found to be 88.7% dense. The estimated purity of the targets was 99.99+% pure and 99.9% pure respectively.

Multiple compositions of (Y,Tb)-123 films were deposited by pulsed laser deposition, using parameters and conditions optimized previously for Y123 [13,14]. The depositions were performed on strontium titanate (STO) and lanthanum aluminate (LAO) substrates with the overall time for film growth being about 20 minutes. Depositions were performed using a Lambda Physik, LPX 300 KrF excimer laser (λ=248 nm). The laser pulse rate was 4 Hz and the laser fluence was ~ 3.2 J/cm^2. The target-to-substrate distance was kept at 6 cm for all of the depositions. The oxygen pressure during the deposition was 300 mTorr for both of the Y,Tb-123 targets, as measured with capacitance manometer and convectron gauges within < 10% variation. Oxygen gas (99.999% purity) flowed into the chamber during growth and the oxygen pressure in the chamber was kept constant using a downstream throttle-valve control on the pumping line. The laser beam was scanned across the targets to improve thickness uniformity of the film. The $LaAlO_3$ (100) and $SrTiO_3$ (100) single crystal substrates were ultrasonically cleaned for 2 minutes, using first acetone followed by isopropyl alcohol. Crystalline substrates were provided by the manufacturer epitaxially polished on both sides of the $LaAlO_3$ and on one side for $SrTiO_3$, and were attached to the heater using a thin layer of colloidal Ag paint. LAO and STO substrates sizes were ~3.2 x 3.2 mm^2 for magnetic J_c measurements.

The background pressure in the chamber was reduced to < 1.4×10^{-4} torr prior to depositions. Samples were heated from room temperature to the deposition temperature of 780 °C at ~1270 °C/h. After deposition, the vacuum pumps and O_2 pressure control were shut off and the films were cooled radiantly from 780 °C to 500 °C while increasing the O_2 pressure to 600 torr. The temperature was then held at 500 °C for 30 minutes. The films were then cooled to room temperature. The (Y,Tb)123 layer thickness was estimated by comparing previous deposition runs in the chamber using the same deposition parameters.

An AC susceptibility technique was used to determine the superconducting transition temperature (T_c). Samples were mounted onto the end of a sapphire rod and measured as the samples were warmed through the transition region at very slow rate of ~ 0.06 K/min. The amplitude of the magnetic sensing field, h, varied from 0.025 Oe to 2.2 Oe, at a frequency of approximately 4 kHz [15,16]. The onset T_c measurements were accurate to within ≤ 0.1 K at three calibration points: liquid He at 4.2 K, liquid N_2 at 77.2 K, and room temperature.

Magnetic J_c measurements were made with a Quantum Design Model 6000 Physical Property Measurement System (PPMS) with a vibrating sample magnetometer (VSM) attachment in fields of 0 to 9 T, and a ramp rate 0.01 T/s. The J_c of the square samples was estimated using a simplified Bean model with $J_c = 30 \, \Delta M/da^3$ where ΔM is in emu, film thickness d and lateral dimension a are in cm and J is in A/cm2. [17]. Samples were subsequently acid-etched at the corners of the samples for thickness measurements. A P-15 Tencor profilometer was used to measure the thickness of the (Y,Tb)123 films. Care was used to measure in twin-free areas of the $LaAlO_3$ substrates, which were observed visually at high magnification. The film thickness and dimensions of each sample were measured multiple times to reduce errors in determination of the superconducting volume and a to < 5 %.

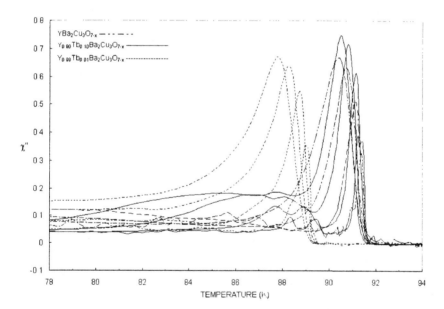

Fig. 1. T_c transitions by AC susceptibility of $Y_{0.9}Tb_{0.1}Ba_2Cu_3O_{7-x}$ and $Y_{0.99}Tb_{0.01}Ba_2Cu_3O_{7-x}$ films compared to $YBa_2Cu_3O_{7-x}$ single layer film with similar thickness. T_cs were measured in 0.025 Oe, 0.25 Oe, 1.0 Oe, and 2.2 Oe applied fields.

RESULTS AND DISCUSSION

The T_c transitions of $Y_{0.9}Tb_{0.1}Ba_2Cu_3O_{7-x}$ and $Y_{0.99}Tb_{0.01}Ba_2Cu_3O_{7-x}$ films are shown in Figure 1, compared to a reference Y123 film. An onset $T_c \sim 91.8$ K was measured for $Y_{0.9}Tb_{0.1}Ba_2Cu_3O_{7-x}$ films, which was similar to a pure Y-123 sample ($T_c \sim 91.9$K). However, in the $Y_{0.99}Tb_{0.01}Ba_2Cu_3O_{7-x}$ films, the T_c was found to be ~ 89.1K. The reason for the decrease of T_c in the 1% Tb-123 films is likely due to the poor quality of commercial Y-123 powder that was used to make the targets. The T_c of undoped Y123 films produced from the targets made using the commercial Y123 powders were found to also be depressed around 88- 89K. It is thought that the impurity phases such as $BaCuO_2$ may be responsible for the reduced T_c in these films. Since the $Y_{0.99}Tb_{0.01}Ba_2Cu_3O_{7-x}$ targets were made using the same commercial Y123 powder, the reduction in T_c is believed to be due to the poor quality of the commercial Y123 powder rather than Tb substitutions. However the $Y_{0.9}Tb_{0.1}Ba_2Cu_3O_{7-x}$ targets were made with in-house produced compositions and are expected to be free from the impurities and hence showed higher T_c. In the future, additional powders will be created avoiding the commercial Y123 for comparison.

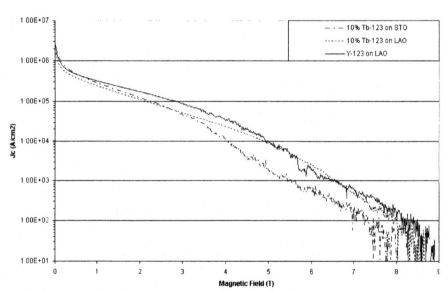

Fig. 2. J_c at 77 K as a function of applied magnetic field for $Y_{0.9}Tb_{0.1}Ba_2Cu_3O_{7-x}$ films made using in-house powders compared to Y-123 films made using in-house Y123 powder.

The magnetic J_cs of the $Y_{0.9}Tb_{0.1}Ba_2Cu_3O_{7-x}$ films are shown in Figures 2 at 77 K, and compared to a reference 123 film made using the same deposition conditions. All powders were made in-house. At zero field, the J_c's of the $Y_{0.9}Tb_{0.1}Ba_2Cu_3O_{7-x}$ samples were between 1.8-2.8 MA/cm^2.

The magnetic J_cs of the $Y_{0.99}Tb_{0.01}Ba_2Cu_3O_{7-x}$ films are shown in Figures 3 at 77 K, and compared to a reference Y123 film made using commercially obtained Y123 powders and using the same deposition conditions. At zero field, the J_c's of the $Y_{0.99}Tb_{0.01}Ba_2Cu_3O_{7-x}$ samples were between 0.6-1.2 MA/cm^2.

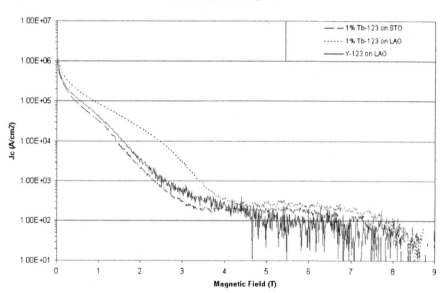

Fig. 3. J_c at 77 K as a function of applied magnetic field for $Y_{0.99}Tb_{0.01}Ba_2Cu_3O_{7-x}$ films made using commercial Y123 powder compared to Y-123 films made using commercial Y123 powder.

CONCLUSIONS

Preliminary results suggest that doping Y-123 with small quantities of Tb-123 (1%) could produce some increase to J_c when subject to fields of 1-4 T. Also, doping with larger quantities of Tb-123 (10%) resulted in little change from Y-123 reference J_c values. The superconducting transition temperature did not reduce significantly for in-house manufactured powders when compared to Y-123.

ACKNOWLEDGEMENTS
The Air Force Office of Scientific Research supported this work.

REFERENCES
[1] P.N. Barnes, G.L. Rhoads, J.C. Tolliver, M.D. Sumption, and K.W. Schmaeman, *Compact, Lightweight Superconducting Power Generators*, IEEE Trans. Mag., accepted.
[2] A. Bourdillon, N. X. Tan Bourdillon, High Temperature Superconductors: Processing and Science (Academic Press, San Diego CA, 1994).
[3] D. Larbalestier, A. Gurevich, D. Matthew Feldmann, A. Polyanskii, Nature 414, 368 (2001).
[4] M. Murakami; D. T. Shaw and S. Jin; Processing and Properties of High T_c Superconductors Volume 1, Bulk Materials; edited by S. Jin, (World Scientific Publishing Co. Pte. Ltd., NJ, 1993).
[5] D. C. Larbalestier and M. P. Maley, MRS Bulletin Aug. (1993) p. 50.
[6] Y. Iijima, K. Onabe, N. Tugaki, N. Tanabe, N. Sadakara, O. Kohno and Y. Ikeno, Appl. Phys. Lett. 60 (1992) 769.
[7] X. D. Wu, S. R. Foltyn, P. N. Arendt, W. R. Blumenthal, I. H. Campbell, J. D. Cotton, J. Y. Coulter, W. L. Hults, M. P. Maley, H. F. Safar and J. L. Smith, Appl. Phys. Lett. 67 (1995) 2397.
[8] A. Goyal, d. P. Norton, J. D. Budai, M. Paranthaman, E. D. Specht, D. M. Kroeger, D. K. Christen, Q. Ile, B. Saffian, F. A. List, D. F. Lee, P. M. Martin, Klabunde, E. Hartfield and V. K. Sikka, Appl. Phys. Lett. 69 (1996) 1795.
[9] S. R. Foltyn, E J. Peterson, J. Y. Coulter, P. N. Arendt, Q. X. Jia, P. C. Dowden, M. P. Maley, X. D Wu, D. E. Peterson, J. Mater. Res. 12(11), 2941 (1997).
[10] T. Aytug, M. Paranthaman, S. Sathyamurthy, B. W. Kang, D. B. Beach, E. D. Specht, D. F. Lee, R. Feenstra, A. Goyal, D. M. Kroeger, K. J. Leonard, P. M. Martin, and D. K. Christen, in ORNL Superconducting Technology Program for Electric Power Systems, Annual Report for FY 2001, p. 1-32 at http://www.ornl.gov/HTSC/htsc.html.
[11] C. R. Fincher, Jr., G. B. Blanchet, *Phys. Rev. Lett.* 67 (1991) 2902.
[12] U. Staub, M. R. Antonio, L. Soderholm, *Phys. Rev. B*, 50, (1994) 7085.
[13] T. J. Haugan, P. N. Barnes, R. M. Nekkanti, I. Maartense, L. B. Brunke, and J. Murphy, Mat. Res. Soc. Symp. Proc. 689E, p. 6.8.1-6.8.5, (2002).
[14] J. W. Ekin, T. M. Larson, N. F. Bergren, A. J. Nelson, A. B. Swartzlander, L. L. Kazmerski, A. J. Panson and B. A. Blankenship, Appl. Phys. Lett. 52 (1988) 1819.
[15] I. Maartense, A. K. Sarkar, and G. Kozlowski, Physica C 181 (1991) 25.
[16] A. Sarkar, B. Kumar, I. Maartense and T. L. Peterson, J. Appl. Phys. 65 (1989) 2392.
[17] J. D. Doss, Engineer's Guide to High-Temperature Superconductivity. New York, John Wiley & Sons Inc., 1989.
M. Murakami; D. T. Shaw and S. Jin; Processing and Properties of High T_c Superconductors Volume 1, Bulk Materials; edited by S. Jin, (World Scientific Publishing Co. Pte. Ltd., NJ, 1993).

FINITE ELEMENT MODELING OF RESIDUAL STRESSES IN MULTILAYERED COATED CONDUCTORS

P. S. Shankar, S. H. Majumdar, S. Majumdar, and J. P. Singh
Energy Technology Division
Argonne National Laboratory
Argonne, IL 60439

ABSTRACT

Finite element analysis (FEA) of a multilayered $YBa_2Cu_3O_{7-x}$ (YBCO)-coated conductor has been performed to assess the applicability of Stoney's equation (which can be used to estimate the stress in a single layer of thin film on a substrate) for residual stress evaluation in multilayered systems. The analysis considered thermal and elastic interactions between the individual layers of the conductor to compute the residual stresses in the layers of the coated conductor. In addition, the changes in substrate curvature that are a result of the deposition of individual layers were also computed, and subsequently these computed curvature changes were used in Stoney's equation to calculate the residual stresses in the various layers of the conductor. The residual stresses thus estimated by Stoney's equation were observed to be in good agreement with those directly computed by FEA, suggesting that Stoney's equation can be used to estimate residual stresses in the various layers of multilayered systems.

INTRODUCTION

$YBa_2Cu_3O_{7-x}$ (YBCO)-coated conductors show great promise for cryogenic applications because of their potential to achieve high critical current densities (J_c) at 77K. Achievement of high J_c in YBCO-coated conductors depends on the successful development of biaxial texture in the YBCO films [1]. Biaxial texture in YBCO films can be achieved by depositing them on textured buffer layers, which can be achieved by processing techniques like inclined substrate deposition (ISD) and ion beam assisted deposition [2-4]. During the processing of coated conductors, residual stresses are developed in various buffer layers and in the superconducting film because of differing thermal expansion properties of the various layers. In addition, intrinsic stresses developed during high-energy processing and layer growth may also contribute to residual stress [5]. These residual stresses may lead to damage and microcracking of the superconducting YBCO film, and thus to degradation of superconducting properties and service life. Therefore, it is critical to evaluate and understand residual stresses in coated conductors to optimize processing for the successful development of high-J_c coated conductors.

Residual stress (σ_f) in a thin film of thickness t_f, deposited on a substrate, can be estimated using Stoney's equation [6], which is given by,

$$\sigma_f = -\frac{E_s t_s^2}{6(1-\nu_s)t_f}\Delta\kappa .$$

(1)

In the above equation, E_s, t_s, and ν_s are the elastic modulus, thickness, and Poisson's ratio of the substrate, and $\Delta\kappa$ is the change in substrate curvature due to the presence of the film. Equation 1 essentially indicates that the stress in the film is proportional to the change of substrate curvature caused by that film. The change in substrate curvature can be measured by techniques such as laser scanning and optical interferometry [7,8]. Stoney [6] originally developed his equation for a thin, single layer of film on a much thicker substrate ($t_f \ll t_s$). For multilayered systems, the use of Stoney's equation is based on the premise that, during deposition of any specific layer, the change in substrate curvature is assumed to be caused only by the specific layer [5]. The equation does not consider the interlayer interactions [5]. Therefore, the stresses estimated by Stoney's equation in multilayered systems may not represent the actual values of stresses present in the specific layers. Thus, to assess the applicability of Stoney's equation for multilayered systems, a three dimensional (3-D) finite element analysis was performed by considering the interaction between the layers and, subsequently, evaluating the residual stresses in the various layers of the coated conductor.

FINITE ELEMENT MODELING

The 3-D finite element analysis (FEA) was performed with ABAQUS standard finite element software. The simulations were performed for an ISD-MgO buffered YBCO-coated conductor of the configuration YBCO/cerium oxide (CeO_2)/yttria-stabilized zirconia (YSZ)/homoepitaxial MgO (HE-MgO)/ISD-MgO/Hastelloy C276 (HC) substrate. For the analysis, the thickness of the HC substrate was taken to be 150 μm. Table I gives the thickness and deposition temperature of the buffer layers and the YBCO film that were used in the FEA. The values of elastic modulus and the coefficient of thermal expansion of the HC substrate, buffer layers, and the YBCO film are given in Table II.

Table I. Thickness and deposition temperatures of various layers of the YBCO-coated conductor used in FEA

Layer	Thickness (μm)	Deposition Temperature (°C)
ISD-MgO	2	Room temperature
HE-MgO	0.5	700
YSZ	0.15	800
CeO$_2$	0.01	800
YBCO	0.3	760

Table II. Mechanical and physical properties of coated conductor materials [9-11]

Material	Elastic Modulus (GPa)	Coefficient of Thermal Expansion (10^{-6}/K)
HC	170-220	12.5
MgO	250	12.8
YSZ	200	9.7
CeO$_2$	200	11.3
YBCO	135-157	13.4

The coated conductor system was modeled as a rectangular geometry (10 mm x 5 mm) and, the simulations were performed so as to replicate the actual fabrication procedure of the multilayered coated conductor. The HC substrate was treated as a 3-D deformable body (stress varying through the thickness) and, all the other layers were considered as planar homogeneous shells, i.e., stress was assumed to be constant within each layer. The effects of intrinsic stresses developed during deposition and stress relaxation due to creep/plastic deformation were ignored in the analysis. Thermal interactions between the layers arising from differences in coefficient of thermal expansion, and elastic interactions due to differences in elastic properties were both considered in the analysis. Residual stresses (at room temperature) in the individual layers of the YBCO-coated conductor were computed in a direction parallel to the plane of the coating layers after deposition of all the layers. Besides computing the residual stresses in the individual layers, finite element simulations were also performed to estimate the change in substrate curvature resulting from the sequential deposition of the individual layers. These computed curvature changes were used in Stoney's equation to evaluate the residual stresses in various layers of the YBCO-coated conductor. A comparison of the FEA-computed residual stresses in the individual layers with those estimated by Stoney's equation (using the computed curvature values) provided guidance on the applicability of Stoney's equation to residual stress evaluation in multilayers.

RESULTS

Figure 1 shows the change in substrate curvature during various deposition steps. The substrate curvature is considered to be zero at room temperature before deposition of any layer. A positive change (convex) in substrate curvature due to the deposition of a layer produces a compressive stress in the layer, whereas a negative change (concave) produces a tensile stress. The data points numbered 1, 4, 7, 10, and 13 in the figure correspond to substrate curvatures at room temperature subsequent to deposition of each layer. Because intrinsic stresses during layer growth have been neglected in the present analysis, the substrate curvature remains constant at each deposition temperature, i.e., at data points 2 and 3, 5 and 6, 8 and 9, and 11 and 12 in the figure. Data point 13 corresponds to the final curvature of the multilayered coated conductor after all the layers have been deposited. Figure 2 shows the residual stresses in the individual layers, calculated by substituting the change in substrate curvature (before and after deposition of a particular layer, Figure 1) into Stoney's equation [6]. For example, the residual tensile stress (≈137 MPa) in the YBCO film was estimated from Equation 1 by using the difference in substrate curvature corresponding to data points 13 and 10 in Figure 1. It should be emphasized that, in using the Stoney's equation, the change in substrate curvature between 13 and 10 is assumed to be caused only by the YBCO layer, whereas in reality, the underlying layers also contribute to the curvature change. Figure 2 also shows the residual stresses at room temperature in the individual layers, computed directly by FEA. It can be clearly seen from the figure that there is little difference between the residual stresses computed by FEA (which considers the interlayer interactions) and those estimated by Stoney's equation using the numerically calculated curvatures. This clearly suggests that interlayer interactions have negligible effects on the development of residual stresses in individual layers of a multilayered system. To confirm the negligible effect of interlayer interactions, the stress in each layer at each stage of deposition was computed by FEA. It was seen that the computed values of stress at room temperature in each individual layer did not significantly vary at various deposition stages. Therefore, it can be concluded that Stoney's equation can be used to estimate residual stresses in specific layers of multilayered coated conductors by measuring the changes in substrate curvature due to the presence of the layers.

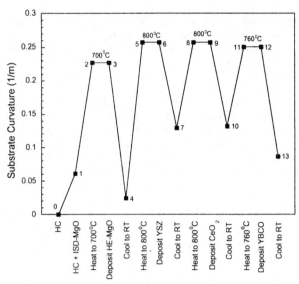

Figure 1. Substrate curvature in YBCO-coated conductor at various deposition steps as estimated by FEA (RT: Room Temperature)

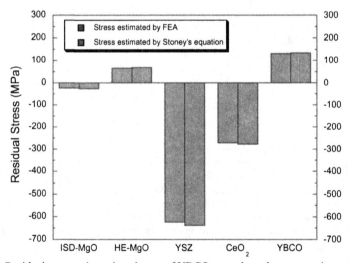

Figure 2. Residual stresses in various layers of YBCO-coated conductor as estimated by FEA and by Stoney's equation (using computed curvatures)

SUMMARY

The applicability of Stoney's equation to residual stress evaluation of individual layers of a multilayered YBCO-coated conductor has been assessed by finite element analysis. The residual stresses in the individual layers of the conductor were evaluated both by FEA and by Stoney's equation (using numerically computed changes in substrate curvature). The values of residual stresses estimated by Stoney's equation were similar to those directly computed by FEA, indicating that Stoney's equation can be used to estimate residual stresses in multilayered coated conductors.

ACKNOWLEDGMENT

This work was supported by the U.S. Department of Energy (DOE), Office of Electric Transmission and Distribution, as part of a DOE program to develop electric power technology, under Contract W-31-109-Eng-38.

REFERENCES

[1] A. Goyal et al., *Appl. Phys. Lett.* **69** 1795 (1996)

[2] K. Hasegawa, K. Fujino, H. Mukai, M. Konishi, K. Hayashi, K. Sato, S. Honjo, Y. Sato, H. Ishii, and Y. Iwata, *Appl. Supercond.* **4** 487 (1996)

[3] Y. Iijima, N. Tanabe, O. Kono, and Y. Ikeno, *Appl. Phys. Lett.* **60** 769 (1992)

[4] B. Ma, M. Li, Y. A. Jee, R. E. Koritala, B. L. Fisher, and U. Balachandran, *Physica C* **366** 270 (2002)

[5] W. D. Nix, *Metall. Trans.* **20A** 2217 (1989)

[6] G. G. Stoney, *Proc. Royal. Soc. Lond. Series A,* **82** 172 (1909)

[7] P. A. Flinn, "Thin Films: Stresses and Mechanical Properties," *Proc. MRS,* **130** 41(1989)

[8] R. E. Cuthress, D. M. Mattox, C. R. Peeples, P. L. Dreike, and K. P. Lamppa, *J. Vac. Sci. Technol. A* **6** 2914 (1988)

[9] A. S. Raynes, S. W. Freiman, F. W. Gayle, and D. L. Kaiser, *J. Appl. Phys.* **70** 5254 (1991)

[10] Private Communication

[11] J. Kawashima, Y. Yamada and I. Hirabayashi, *Physica C,* **306** 114 (1998)

PULSED LASER DEPOSITION OF Nd DOPED $YBa_2Cu_3O_{7-\delta}$ FILMS

J. C. Tolliver, T. J. Haugan, S. Sathiraju, N. A. Yust, P. N. Barnes
Air Force Research Laboratory
1950 5th St., Bldg. 18
Wright-Patterson AFB, OH 45433

C. Varanasi
University of Dayton Research Institute
300 College Park
Dayton, OH 45469

ABSTRACT
    In this study, experiments were conducted to incorporate Nd into $YBa_2Cu_3O_{7-\delta}$ (YBCO) thin films through pulsed laser ablation of a $(Y_{0.6}Nd_{0.4})Ba_2Cu_3O_{7-\delta}$ target. The processing conditions to get high T_c films were found to deviate from the standard conditions that are generally used for undoped YBCO films and so considerable process optimization was required. Characterization results of the best films obtained so far are presented.

INTRODUCTION
    Flux pinning enhancement in YBCO coated conductors is of great interest to further advance the critical current density (J_c) at high magnetic fields for applications. Future applications such as superconducting generators, motors, etc. require that these coated conductors operate at such fields. Several approaches can be considered to enhance J_c in YBCO films viz. precipitate incorporation [1], stress field induced flux pinning through rare earth ion substitution [2] or combination of both [3]. Flux pinning enhancement through rare earth ion substitution in $YBa_2Cu_3O_{7-\delta}$ (YBCO) has been demonstrated in powders [4] and melt processed materials [5]. Flux Pinning due to rare earth ion substitutions in YBCO films is presently being investigated by several groups [6, 7]. In the present study Nd substituted YBCO films were processed by pulsed laser ablation to investigate the potential J_c enhancements and the characterization results of the films is presented.

EXPERIMENTAL
    Superconducting YBCO and $NdBa_2Cu_3O_{7-\delta}$ (NdBCO) powders were prepared via the standard solid reaction method using calculated amounts of yttrium oxide, neodymium oxide, copper oxide, and barium carbonates. A 1" diameter target was pressed using YBCO and NdBCO powders mixed in a nominal composition of $(Y_{0.6}Nd_{0.4})Ba_2Cu_3O_{7-\delta}$ and the target was sintered at 910°C in air and 860°C in 1% oxygen. Heat treatment with intermediate grinding cycle was repeated several times till a 92% dense, homogeneously mixed target was obtained.
    Pulsed laser deposition was accomplished using a Lambda Physik LPX 305i KrF excimer laser (λ=248 nm) in a vacuum chamber manufactured by Neocera, Inc. Single crystal lanthanum aluminate (LAO) and strontium titanate (STO) substrates were heated

by contact heating method on a heater block using colloidal silver paste. Target to substrate distance was maintained at 6 cm for all the experiments. Substrate temperature was measured using a thermocouple mounted in the heater block. Deposition pressure was maintained using a downstream pressure control valve and a low oxygen gas flow. Laser energy was kept constant and a fluence of approximately 4.2 J/cm^2 was measured. The laser beam was allowed to scan on the target during the deposition to provide more uniform ablation. Several films were deposited by varying the deposition conditions such as temperatures from 730-850°C and oxygen pressures from 300-450 mTorr.

The critical transition temperature (T_c) of the films was measured by ac susceptibility in varying magnetic fields from 0.025 to 2.2 Oe as well as by four probe transport current method using a constant current of 100 μA. Microstructure of the films was observed in a scanning electron microscope. Patterning was done on the films to perform the transport critical current density (J_c) measurement. The bridge width was maintained at 0.5 mm and the thickness of the film was measured by a KLA Tencor P-15 surface profiler and was found to be approximately 0.25 μm. The transport J_c measurements were conducted at 77 K in self-field using 1 μV/cm criteria.

RESULTS & DISCUSSIONS

Films processed at a deposition temperature of 735°C and 400 mTorr of oxygen pressure and at 750°C and 300 mTorr of oxygen pressure yielded films with T_c higher than 90 K. Other films with different deposition conditions of varying temperature and oxygen pressures yielded films with scattered T_c ranging anywhere from 82-90 K. Generally with the similar target to substrate distance and laser fluence, high quality undoped YBCO films are routinely grown at 780°C and 300 mTorr of oxygen pressure in this chamber. However, to get high quality Nd doped films in this study, considerable optimization was required.

Due to the presence of different rare earth metals Yttrium and Neodymium in the present target, the plume dynamics are expected to be different from the plumes generated from an undoped YBCO target that contained only Yttrium. Since Y and Nd have different ionic radii (Nd^{3+} = 0.995 Å, Y^{3+} = 0.893 Å) and atomic weight (Nd=144.24 g/mol, Y= 88.90 g/mol), the transport velocities of these ions to the growing film are expected to be different. Also in the film, substitutions of Nd in Ba sites are possible since ionic sizes are sufficiently similar (Nd^{3+} = 0.995 Å, Ba^{2+} = 1.34 Å); this may explain the reduced T_c of the Nd doped films. The extent of these substitutions can depend upon the deposition conditions such as oxygen pressure and temperature which can explain why different T_c values were observed in films grown at different deposition conditions. All of these mechanisms may be responsible for variations in film quality requiring substantial variance in the deposition conditions as compared to the conditions used for undoped YBCO to get high T_c films. Presently, the characterization of the films for Nd occupancy in Y or Ba sites is being investigated to understand the variations in the T_c of the films grown at different conditions. As such, it is not clear at this point what the final Y to Nd ratio was in the final films pending additional characterization

Figure 1 shows the ac susceptibility curves for one of the films grown at optimized conditions. It can be seen that the onset temperature is around 90 K. As the magnetic field is increased, the broadening of the curves is minimal indicating the presence of a high quality film with reasonably good J_c [8]. Also, there is no indication of alternate regions of the film with different T_cs as indicated by the absence of "kinks" in the susceptibility curve at various transition temperatures. Figure 2 shows the four-probe resistivity measurement of the film indicating that the onset T_c is around 90 K, agreeing well with the ac susceptibility data. Even so, the width of the transition is somewhat broader than the sharper transitions of plain YBCO.

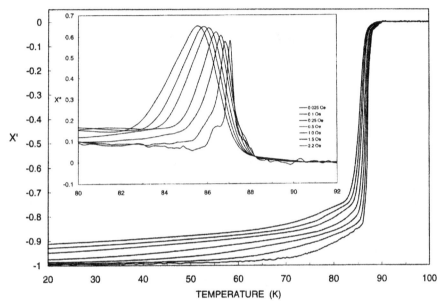

Figure 1. AC Susceptibility curves at different magnetic fields of a film grown using a $(Y_{0.6}Nd_{0.4})Ba_2Cu_3O_{7-\delta}$ target

Figure 3 shows a scanning electron micrograph of a high T_c film grown using Nd doped YBCO target. It can be seen that the grain size of the films is roughly 0.5 μm and that the films are dense with no apparent cracking or such defects. Presence of larger particles was also observed but the composition of these particles is yet to be determined. Since the targets are 92% dense, it is plausible that some of the particles were created during laser ablation of the target and ejected to the film. Figure 4 shows the transport current measurement data taken at 77 K in self-field on a 0.5 mm wide bridge for a 0.25 μm thick film. This particular sample carried a transport current of 3.7 A, resulting in a J_c of 3.1×10^6 A/cm^2.

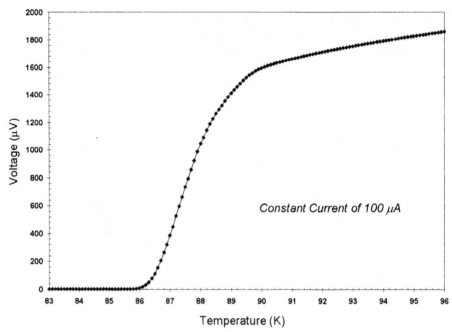

Figure 2. Resistivity measurement of a film grown from a $(Y_{0.6}Nd_{0.4})Ba_2Cu_3O_{7-\delta}$ target

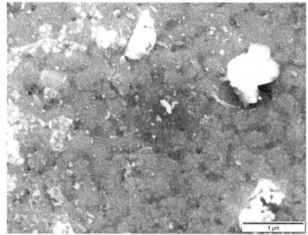

Figure 3. Scanning electron micrograph of a Nd doped YBCO film

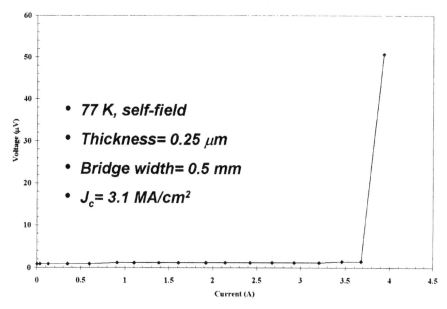

Figure 4. Transport J_c of a film measured at 77K in self-field

CONCLUSIONS

Nd doped YBCO films have been produced using a $(Y_{0.6}Nd_{0.4})Ba_2Cu_3O_{7-\delta}$ target. T_c of the films varied depending on the processing conditions and was found to be much more sensitive to the temperature and oxygen pressure as compared to undoped YBCO. After optimizing of the deposition conditions, good quality films with a T_c of 90 K were obtained. A transport J_c of over 3 MA/cm^2 has been achieved in an Nd-substituted YBCO thin film.

REFERENCES

[1]T.J. Haugan, P.N. Barnes, I. Maartense, E.J. Lee, M.D. Sumption and C.B. Cobb, *J. Mater. Res.*, **18** 2618 (2003).

[2]H.H. Wen, Z.X..Zhao, R.L. Wang, H.C. Li, B.Yin, *Physica C*, **262** 81-88 (1996).

[3]C. Varanasi, R. Biggers, I. Maartense, D. Dempsey, T.L. Peterson, J. Solomon, J. McDaniel, G. Kozlowski, R. Nekkanti, C.E. Oberly, "Advances in Laser Ablation of Materials Symposium," Mater.Res.Soc, Warrendale, PA, Edited by Singh RK, DH Lowndes, Chrisey DB, Fogarassy E, Narayan J 263-8 (1998).

[4]S. Jin, T.H. Tiefel, G.W. Kammlott, R.A. Fastnacht, J.E. Graebner, *Physica C* **173** 75-79 (1991).

[5]M.R. Koblischka, M.Muralidhar, Masato Murakami, *Physica C* **337** 31-38 (2000).

[6]A. Radhika Devi. V. Seshu Bai, P.V. Patanjali, R. Pinto, N. Harish Kumar, and S K Malik, *Supercond. Sci. Technol.*, **13** 935-939 (2000).

[7]Yang Li, Zhong-Xian Zhao, *Physica C* **351** 1-4 (2001).

[8]P.N. Barnes, I. Maartense, T.L. Peterson, T.J. Haugan, A.L. Westerfield, L.B. Brunke, S. Sathiraju, and J.C. Tolliver, *Mat. Res. Soc. Symp. Proc.*, EXS-3, pp. EE6.4.1-3 (2004).

Buffer Layers

EPITAXIAL GROWTH OF Eu$_3$NbO$_7$ BUFFER LAYERS ON BIAXIALLY TEXTURED Ni-W SUBSTRATES

M.S. Bhuiyan, M. Paranthaman, D. Beach, L. Heatherly, A. Goyal, and E.A. Payzant
Oak Ridge National Laboratory
Oak Ridge, TN 37831, USA

K. Salama
University of Houston
Houston, TX 77204, USA

ABSTRACT

Epitaxial film of Eu$_3$NbO$_7$ was deposited on biaxially textured nickel-3 at.% tungsten (Ni-W) substrates by using sol-gel processing . Precursor solution of 0.50 M concentration of total cation was spin coated on short samples of Ni-W substrates and films were crystallized at 1050°C in a gas mixture of Ar-4%H$_2$ for 15 minutes. High temperature *in situ* x-ray diffraction (HTXRD) studies show that the nucleation of Eu$_3$NbO$_7$ films starts at 800 °C. $\theta/2\theta$ x-ray diffractograms revealed only (004) reflections, indicating a high degree of out-of-plane texture. Detailed X-Ray studies indicate that Eu$_3$NbO$_7$ films has good out-of-plane and in-plane textures with full-width-half-maximum values of 6.8° and 8.21°, respectively. Scanning electron microscopy showed that the films were smooth, continuous, and free of pin holes. Efforts are under way to grow YBCO films on sol-gel derived Eu$_3$NbO$_7$ buffer layers.

INTRODUCTION

As an alternative to the relatively higher cost and complexity of vacuum deposition techniques, chemical solution processing techniques have emerged as viable low-cost nonvacuum methods for producing ceramic oxide powders and films [1-3]. These processes offer many desirable aspects, such as precise control of metal oxide precursor stoichiometry and composition, ease of formation of epitaxial oxides, relatively easy scale-up of the film and possibly low cost. In the Rolling-Assisted Bi-axially Textured Substrates (RABiTS) approach, three-layer architecture of CeO$_2$/YSZ/Y$_2$O$_3$/Ni-W is used to fabricate long lengths of buffered tapes. The purpose of the buffer layers is to retard oxidation of Ni, to reduce the lattice mismatch between Ni and YBCO and also to prevent diffusion of Ni into YBCO.

We have been examining solution routes to epitaxial buffers on Ni-W substrates to simplify the multilayer buffered architecture to single layer buffered architecture. In recent years various rare-earth oxides (RE$_2$O$_3$) and rare-earth zirconium oxide (RE$_2$Zr$_2$O$_7$) films have been grown epitaxially on biaxially textured Ni and Ni-W substrates by solution based methods [4-8]. A new series of single rare earth niobate, Gd$_3$NbO$_7$ buffer layers [9] has been reported for the growth of superconducting YBa$_2$Cu$_3$O$_{7-\delta}$ (YBCO) films on biaxially textured Ni-W (3 at.%) substrates. Preliminary results show that YBCO films with a J$_c$ of 1.1 MA/cm^2 can be grown on Gd$_3$NbO$_7$ buffered Ni-W substrates. Here we report in detail about the nucleation and growth of Eu$_3$NbO$_7$ (ENO) on Ni-W substrates. Eu$_3$NbO$_7$ has a cubic pyrochlore structure with a lattice parameter of 10.672 Å (pseudo cubic lattice parameter: 3.773 Å) and it has a good chemical compatibility with Ni-W and also good thermal stability. In this paper, we describe our successful development of the growth of Eu$_3$NbO$_7$ buffer layer on rolled Ni-W substrates by sol-gel method.

EXPERIMENTAL DETAILS

All solution manipulations were carried out under an atmosphere of argon using standard Schlenk techniques. Niobium ethoxide and 2-methoxyethanol (Alfa) were used as received and europium acetate (Alfa) was purified beforehand. Europium acetate was prepared by the reaction of europium oxide (Alfa, 99.99%) with a 5-fold excess of 25% acetic acid at 80 °C for one hour [10]. The resulting clear solution was filtered and the solvent removed. The precipitate was dried at 120 °C under dynamic vacuum for 16 hours. A small sample was allowed to react with excess water and then dried to constant weight at 35 °C to form the well-characterized tetrahydrate. The weight gain of this sample allowed us to estimate that the degree of hydration of the europium acetate after vacuum drying is 1.5 ± 0.2.

Europium methoxyethoxide solution in 2-methoxyethanol was prepared by charging a flask with 1.234 g (3.75 m moles) of europium acetate and 10 ml of 2-methoxyethanol. The flask was refluxed for 1 h at 130°C, and 0.398 g (1.25 m moles) of niobium ethoxide was added. The contents were rediluted with additional 2-methoxyethanol, and the distillation/redilution cycle was repeated twice more to completely exchange the metal alkoxide ligand for the methoxyethoxide ligand. The final concentration was adjusted to produce 10 ml of a 0.50 M stock solution. This solution was then spin coated onto short cube textured Ni-W substrates of 2 cm x 1 cm in size at 5000 rpm for 30 sec; followed by heat treatment at 1050°C for 15 min. in a reducing atmosphere of Ar-4% H_2. The samples were introduced into a pre-heated furnace kept at 1050°C after a 5 minutes purge with Ar-4% H_2 gas mixture at room temperature. After 15 min. heat-treatment at 1050°C, the samples were quenched to room temperature with the same atmosphere. The heating and cooling rates were in the range of 350–400°C/min. About 20 nm thick Eu_3NbO_7 films were produced in a single coat and multiple coatings were made to prepare thick films.

The Eu_3NbO_7 films were characterized by using x-ray diffraction (XRD) for phase purity and texture, high temperature XRD for nucleation, scanning electron microscopy (SEM) for homogeneity and microstructure. A Philips model XRG3100 diffractometer with CuKα radiation was used to record the θ–2θ XRD patterns. The texture analysis was performed using a Picker 4-circle diffractometer. High temperature *in situ* XRD experiments were carried out in a flowing atmosphere of He–4% H_2 and heating ramp of 400°C/min on a Scintag PAD X diffractometer with an mBraun linear position sensitive detector (PSD) covering a 8 ° range centered at $2\theta = 31$ °. The microstructure analyses of these samples were performed by using a Hitachi S-4100 field emission SEM.

RESULTS AND DISCUSSION

For *in-situ* high temperature x-ray diffraction the sample was heated from room temperature to 1200°C at a heating rate of 400°C/min in a reducing atmosphere of He–4%H_2 and the θ–2θ XRD patterns were recorded at 400, 600, 800, 900, 1000, 1100, and 1150°C. The experimental set-up for the high temperature XRD is shown in figure 1. Plot for nucleation of Eu_3NbO_7 film on textured Ni–W substrates is shown in figure 2, which indicates the nucleation of the film around 800°C.

A typical θ–2θ XRD scan for a spin-coated Eu_3NbO_7 film on the Ni–W substrate is shown in figure 3. The intense Eu_3NbO_7 (004) peak reveals the presence of a c-axis-aligned film. The typical (222) pole figure in log scale for a Eu_3NbO_7 film grown on the Ni–W substrate is shown in figure 4, which indicates the presence of a single cube-on-cube texture. The ω (out-of-plane) and φ (in-plane) scans of these films on the Ni–W substrates are shown in figure 5.

The Eu_3NbO_7 film has a good out-of-plane and in-plane texture with full-width at-half-maximum (FWHM) of 6.8° and 8.21°, respectively, as compared to the Ni–W substrates values of $\Delta\omega$ = 5.51° and $\Delta\varphi$ = 6.72°.

SEM micrographs for Eu_3NbO_7 buffer layers on rolled Ni-W substrate using spin coating are shown in figure. 6. As seen from figure 6, Eu_3NbO_7 buffer layers provide very good coverage for the Ni-W surface. Most of the Ni-W grain boundary grooves on the Ni-W surface were found to be well covered. Figure 6 also shows that the buffer layers are continuous as well as crack free.

These results indicate sol-gel techniques can produce continuous, dense and crack-free buffer layers on rolled Ni-W substrate. Efforts are being made to deposit YBCO by pulsed laser deposition (PLD) on Eu_3NbO_7 (sol-gel)/Ni-W.

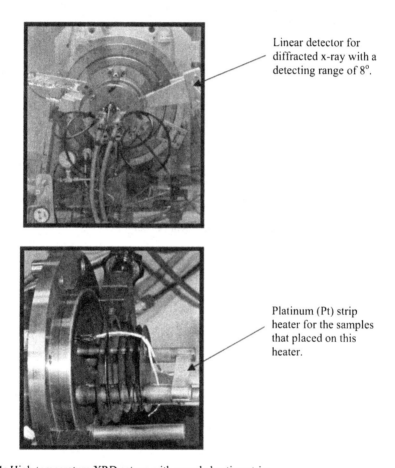

Linear detector for diffracted x-ray with a detecting range of 8°.

Platinum (Pt) strip heater for the samples that placed on this heater.

Fig.1: High temperature XRD set-up with sample heating strip

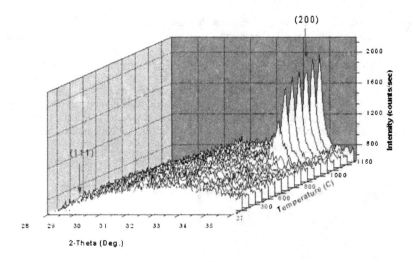

Fig. 2: A typical θ–2θ scan obtained for a 20 nm thick Eu_3NbO_7 buffered Ni–W substrates in a high temperature *in situ* XRD heat-treated at various temperatures. The Eu_3NbO_7 film has a preferred c-axis orientation and the crystallization starts around 800 °C

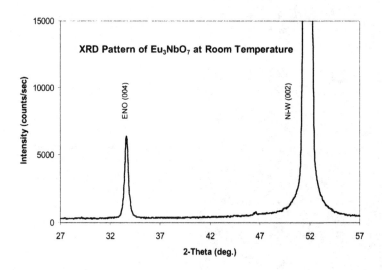

Fig. 3: A typical room-temperature θ–2θ scan obtained for a 20 nm thick Eu_3NbO_7 buffered Ni–W substrates. The Eu_3NbO_7 film has a preferred c-axis orientation.

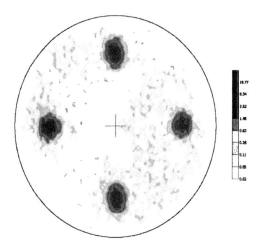

Fig. 4: The typical Eu$_3$NbO$_7$ (222) pole figure obtained for a 20 nm thick film grown on the textured Ni–W substrate.

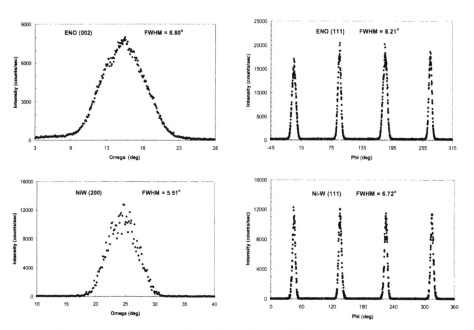

Fig. 5: The ω and φ scans obtained for a 20 nm thick Eu$_3$NbO$_7$ film grown on textured Ni–W substrate. The FWHM values for each scan are shown inside the patterns.

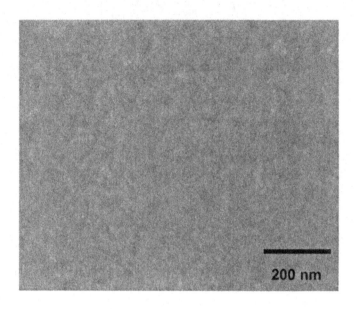

Fig. 6: SEM micrograph of the film surface deposited on Ni-W.

SUMMARY
We have successfully developed a new chemical solution process to grow epitaxial Eu_3NbO_7 buffer layers on Ni-W (002) substrates. In-plane and out-of-plane alignments indicate the buffer layers are sharp textured. Pole figures of buffer layers show a predominantly cube-on-cube texture. SEM micrograph reveals a continuous, dense and crack-free microstructure of the solution deposited buffers which indicate that the surface of Eu_3NbO_7 buffer layers deposited on rolled Ni-W substrates using sol-gel process is suitable for deposition of YBCO film.

ACKNOWLEDGEMENTS
This work was sponsored by the United States Department of Energy, Office of Science, Division of Materials Science, Office of Electric Transmission and Distribution - Superconductivity program. This research was performed at the Oak Ridge National Laboratory, managed by U.T.-Battelle, LLC for the USDOE under contract DE-AC05-00OR22725. M. S. Bhuiyan would also like to acknowledge the help of AFOSR for providing a financial support.

REFERENCES
[1] M.W. Rupich, Y.P. Liu, and J. Ibechem., Appl. Phys. Lett **60**, 1384 (1992).
[2] M. Paranthaman, and D.B. Beach, J. Amer. Ceram. Soc. **78**, 2551 (1995).
[3] F.F. Lange, Science 273, 903 (1996).

[4] M. Paranthaman, T.G. Chirayil, F.A. List, X. Cui, A. Goyal, D.F. Lee, E.D. Specht, P.M. Martin, R.K. Williams, D.M. Kroeger, J.S. Morrel, D.B. Beach, R. Feenstra, and D.K. Christen, J. Amer. Ceram. Soc. **84**, 273 (2001).

[5] S. Sathyamurthy, M. Paranthaman, T. Aytug, B.W. Kang, P.M. Martin, A. Goyal, D.M. Kroeger, and D.K. Christen, J. Mater. Res. **17**, 1543 (2002).

[6] T. Aytug, M. Paranthaman, B.W. Kang, D.B. Beach, S. Sathyamurthy, E.D. Specht, D.F. Lee, R. Feenstra, A. Goyal, D.M. Kroeger, K.J. Leonard, P.M. Martin, and D.K. Christen, J. Amer. Ceram. Soc. **86**, 257 (2003).

[7] M.S. Bhuiyan, M. Paranthaman, S. Sathyamurthy, T. Aytug, S. Kang, D.F. Lee, A. Goyal, E.A. Payzant, and K. Salama, Supercond. Sci. Technol. **16**, 1305 (2003).

[8] M.S. Bhuiyan, M. Paranthaman, S. Sathyamurthy, T. Aytug, S. Kang, D.F. Lee, A. Goyal, E.A. Payzant, and K. Salama, Fall Mater. Res. Soc. Symp. Proc. **EXS-3**, EE4.5.1 (2004).

[9] M. Paranthaman, M.S. Bhuiyan, S. Sathyamurthy, H.Y. Zhai, A. Goyal, and K. Salama, J. Mater. Res. (submitted).

[10]L. Gmelin, "Gmelin Handbook of Inorganic Chemistry", 8th Ed., Springer-Verlag, Berlin, p. 34 (1984).

PULSED LASER DEPOSITION OF $(Y_{1-x}Ca_x)Ba_2NbO_6$ (x = 0.0, 0.05, 0.1, 0.2, 0.4) BUFFER LAYERS

S. Sathiraju, G. A. Levin
National Research Council
PRPG,Air Force Research Laboratory
Dayton, OH 45433

J.P. Murphy
University of Dayton Research Institute
PRPG, Air Force Research Laboratory
Dayton, OH 45433

I. Maartense,
MLPS, Materials Laboratory
Air Force Research Laboratory
Dayton, OH-45433

T. L. Peterson
MLPS, Materials Laboratory
Air Force Research Laboratory
Dayton, OH-45433

T. Campbell, J.C. Tolliver, T. J. Haugan and Paul N. Barnes
Propulsion Directorate, Air Force Research Laboratory,
Wright-Patterson Air Force Base, OH 45433

ABSTRACT

The growth optimization of $Y_{1-x}Ca_xBa_2NbO_6$ (YCBNO), x = 0.0 ,0.1.0.2,0.4, thin films on LaAlO₃ (100) and MgO (100) single crystals as well as ion beam assisted deposition (IBAD) of MgO buffered Hastelloy substrates has been investigated using pulsed laser deposition. X-ray diffraction confirms the epitaxial growth of highly *400* oriented YBNO thin films on single crystal substrates and IBAD MgO buffered Inconel substrates for x = 0.0 and 0.1. For x > 0.1 Ca doping, *220* oriented growth dominated. Atomic force microscopy studies of the surface morphology revealed that the best average surface roughness of the YBNO films deposited on buffered substrates is 2nm, and with increasing Ca content the surface roughness increases considerably. The critical temperature (T_c) of $YBa_2Cu_3O_{7-x}$ (Y-123) thin films deposited on these YBNO buffer layers ranges from 80 to 87 K. The results presented here are preliminary in nature.

INTRODUCTION

Recent advancements in high temperature superconducting (HTS) coated conductor research based on the ion beam assisted deposition (IBAD) architecture show promise for the textured MgO IBAD buffer. However, it has been reported that for Y123 films deposited on MgO substrate, an interlayer of barium salt is formed at the interface if the processing temperature is above 700 °C[1]. Also, another problem reported with the films grown on MgO is the presence of 45° in-plane rotated grains which degrades the crystalline quality of the deposited $RBa_2Cu_3O_x$ (where R = Y, rare earths)[2,3]. A possible solution is to use a suitable buffer as an interlayer between the MgO layer and the subsequent Y123 superconductor films. Initial results of YBa_2NbO_6 (YBN) thin layers on IBAD MgO architecture are promising[4]. Effects of Ca doping in Y123 films has been extensively studied because of the opportunity to explore the regime of over doped charge carriers given by the Ca $^{+2}$ substitution of Y^{+3} cation[5-7]. Also it is well known that transport properties of Ca doped Y-123 films dramatically improved[5]. In fact, it is well known that GBs in YBCO superconductors are depleted of carriers leading to a reduced intergrain critical current density[6]. Moreover, it was shown that is possible to restrict the Ca doping to the YBCO GBs by using $Y_{1-x}Ca_xBa_2Cu_3 O_y$ /Y-123 multi-layers[7]. We have already investigated the growth of YBNO buffer layers for coated conductor applications[4]. As such, it is of interest to investigate the Ca doping at the Y site of YBNO perovskite oxide dielectric films. In this paper, we report the effect of Ca doping in YBNO thin films.

EXPERIMENTAL

In the present study, the films were prepared by the pulsed laser deposition (PLD) using a Lambda Physik 304i excimer laser using KrF at the 248 nm wavelength. We have used in-house prepared stoichiometric sintered $Y_{1-x}Ca_xBa_2NbO_x$ (YCBNO) targets. The targets were prepared by following conventional solid state reaction processing. High purity (99.99%) Y_2O_3, $CaCO_3$, $BaCO_3$, and Nb_2O_5 powders were thoroughly mixed in the stoichiometric ratio for x = 0.0, 0.1, 0.2 and 0.4. The mixture was then calcified at $1350°$ C for 48 hrs with three intermediate grindings. The phase purity of the calcified material was checked by x-ray diffraction and the phase pure material was finely grounded and palletized at a pressure of 3 Tons in the form of circular discs having 1 inch diameter and 3 mm thickness. These discs were sintered again at $1400°C$ for 24 hrs. All processing was carried out in the air. The MgO(100) single crystals were commercially obtained and the IBAD MgO textured substrates used in this study were fabricated by Superconductivity Technological Center, Los Alamos National Laboratory [8]. The dimensions of the substrates used in this process were 10mm x 10mm x 0.25mm. The substrates were adhered to the substrate heater using silver paste.

Table1. Deposition conditions

Parameters	$Y_{1-x} Ca_xBa_2NbO_y$ X = 0.0, 0.1, 0.2, 0.4	YBa_2Cu_3Oy
Deposition Temperature	850 °C	750-820 °C
Oxygen Pressure	250 mTorr	230mTorr
Laser Energy	2.5 J/cm2	2 J/Cm2
Laser frequency	10 Hz	4 Hz
Substrate-Target distance	6 cm	6 cm
Substrates	MgO, IBAD MgO	YCBNO buffered substrates
Thickness	300 nm	2000 nm

The base pressure obtained for the deposition chamber was 10^{-7} Torr and initial depositions were carried out using a reduced atmosphere. Then chamber was back filled with O_2 gas of 200-300 m Torr and the deposition of YCBNO was performed for 10 minutes. Details of deposition parameters used are summarized in Table 1. After the deposition, films were cooled to room temperature in 500 Torr of oxygen. Y-123 films were subsequently deposited on these buffer layers using PLD (Table 1).

The as deposited films were analyzed by detailed x-ray diffraction studies. Two-theta (2θ) scans were accomplished by using a Rigaku x-ray diffractometer. A Philips MRD with four circle diffractometry was used to study the crystalline alignment of the substrate, buffer layers, and the superconductor by means of phi scans. The microstructure of the various films was evaluated using atomic force microscopy (AFM) with particular emphasis on surface roughness and surface morphology of the films for MgO and IBAD MgO buffered substrates. Y-123 films were deposited using PLD over the applied buffer layers. AC susceptibility and transport current measurements were performed to estimate the critical temperature (T_c) and critical current density (J_c) of the films.

RESULTS AND DISCUSSION

Figure 1a and 1b shows the XRD of the YBNO (x = 0.0) films deposited on MgO single crystal and IBAD MgO substrate. YBNO films deposited on MgO substrate is highly (*400*) oriented. However, weak reflections of (*220*) were observed around 30°. In the case of films grown on IBAD MgO substrate, weak (*220*) and strong (*400*) reflections were observed. Films were deposited on both substrates at the same time at 850°C. Figure 3 shows the XRD of Y-123 film

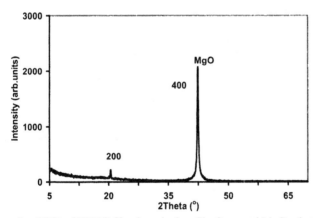

Figure1a. XRD of YBNO film deposited on Single crystal MgO substrate

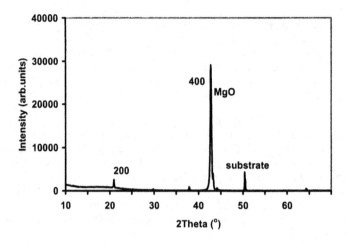

Figure1b. XRD of YBNO film deposited on IBAD MgO buffered Hastelloy substrate

deposited on YBNO buffered IBAD MgO/Hastelloy substrate. We have observed highly c-axis oriented Y-123 growth on YBNO buffered IBAD MgO buffered Hastelloy substrate.

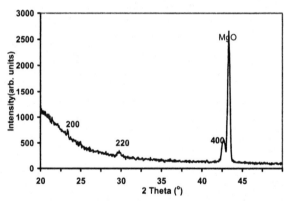

Figure 2. XRD of $Y_{0.9}Ca_{0.1}Ba_2NbO_6$ film deposited on IBAD MgO buffered Hastelloy substrate

Figure 2 shows the XRD of the YCBNO film for x = 0.1 on IBAD MgO buffered Hastelloy substrate. We have observed weak (*200*) and (*220*) and strong (*400*) reflections for x = 0.1 concentration. Broadening of the YCBNO peaks is clearly seen from the XRD patterns. Figure 3 shows the XRD pattern of YCBNO film for x = 0.2 deposited on IBAD MgO buffered metal substrate. Decrease of the (*400*) oriented peak and increase of (*220*) peak were observed when Ca concentration is increased by 20% .

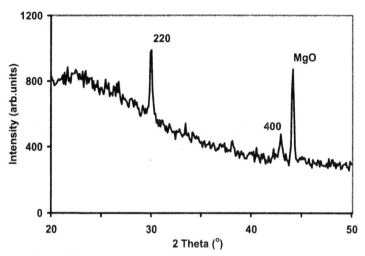

Figure 3 XRD pattern of YCBNO film for x = 0.2 concentration.

When x = 0.4 for the Ca doped concentration, it was observed that (220) orientation completely dominates the growth (Figure 4). We have not seen any (400) oriented peak or (200) oriented peak in the XRD pattern.

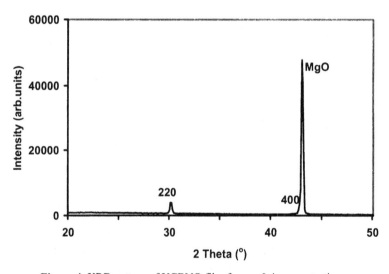

Figure 4. XRD pattern of YCBNO film for x = 0.4 concentration

Overall, the Ca concentration affected the growth orientation, although no major changes in the structure of YBNO films were observed. Surface roughness is estimated after measuring 30 points on the surface of the film using atomic force microscope. The average surface R_a is 1.5 nm. Table 2 shows the average surface roughness of YCBNO films deposited on IBAD MgO buffered Hostelry substrates.

Table 2 Surface roughness analysis of YCBNO films

$Y_{1-x}Ca_x Ba_2 NbO_6$ Film	Average R_a (nm)
X = 0	1.5
X = 0.1	2.5
X = 0.2	3.0
X = 0.4	6.0
YBNO/I-MgO	1.5
YBNO/MgO	6

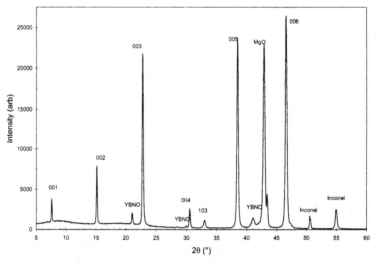

Figure 5. X-ray pattern of Y-123 film deposited on YBNO/IBAD-MgO/Hastelloy substrate

Figure 5 shows the XRD pattern of Y-123 deposited on YBNO/IBAD-MgO/Hastelloy. It is clear from the figure that both YBNO and Y-123 have their a-b planes aligned exclusively along the MgO high symmetry direction as evident from the appearance of diffraction peaks with $90°$ spacing apart. The epitaxial relationship exists between the different films and the IBAD MgO buffer $[100]_{MgO}$ parallel $[100]_{YBNO}$, $[001]_{YBNO}$ parallel $[001]_{Y-123}$

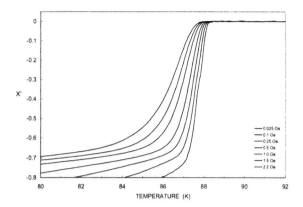

Figure 6. A.C. Susceptibility of Y-123 film deposited on YBNO Buffered IBAD
MgO/Hastelloy

Figure 6 shows the AC susceptibility curves of Y-123 thin film deposited on YBNO
buffered IBAD MgO deposited Hastelloy substrate. The magnetic field of the susceptometer
was varied between 0.0 - 2.2 Oersted. The depressed T_c from typical Y123 films suggests there is
potential for additional optimization of the growth parameters. The critical temperature (T_c) of
$YBa_2Cu_3O_{7-x}$ (Y-123) thin films deposited on these YBNO buffer layers ranged from 80 to 87 K.
Note that the deposition of YCBNO may be further optimized since the IBAD MgO layers were
only of sufficient quality to test for compatibility and epitaxial growth of the new buffer. Hence,
the results presented here are preliminary in nature and can be improved upon.

CONCLUSIONS

The effect of Ca doping at the Y site in YBNO films has been investigated. $Y_{1-x}Ca_xBa_2NbO_6$ (YCBNO) films for x = 0.0, 0.1, 0.2, and 0.4 concentrations were deposited by
PLD on MgO (100) single crystals and ion beam assisted deposition (IBAD) MgO buffered
Hastelloy substrates. X-ray diffraction confirms the epitaxial growth of highly (*400*) oriented
YCBNO thin films on single crystal substrates and IBAD MgO buffered Inconel substrates for x
= 0.0 and 0.1. The atomic force microscopy of surface morphology studies revealed that the best
average surface roughness of the YCBNO films deposited on buffered substrates is 2 nm for x =
0. Increasing the Ca doping concentration tends to favor (*220*) orientations and also resulted in
the increase of surface roughness. The critical temperature (T_c) of $YBa_2Cu_3O_{7-x}$ (Y-123) thin
films deposited on these YBNO buffer layers ranges from 80 to 87 K. The results presented here
are preliminary in nature.

ACKNOWLEDGEMENTS

Sathiraju is thankful to the National Academy of Sciences National Research Council, and
Propulsion Directorate of the Air Force Research Laboratory for the Sr. Research Fellowship and

for the support of his research on coated conductors. The authors thank A. Campbell for providing AFM of the samples, and P. Arendt and Quanxi Jia for the IBAD MgO sample.

REFERENCES

[1]Koinuma et al., *Jpn. J. Appl. Phys.* **27,** L1216 (1988)

[2]Mukaida, M; Takano, Y; Chiba, K; Moriya, T; Kusunoki, M; Ohshima, S "A new epitaxial $BaSnO_3$ buffer layer for $YBa_2Cu_3O_{7-[delta]}$ thin films on MgO substrates" Supercond. Sci. Tech. **12**, 890 (1999)

[3]J.R. Groves, P.N. Arendt, S.R. Foltyn, Q.X. Jia, T.G. Holesinger, H. Kung, E.J. Peterson, R.F. DePaula, P.C. Dowden, L. Stan, L.A. Emmert, "High critical current density $YBa_2Cu_3O_{7-\delta}$ thick films using ion beam assisted deposition MgO bi-axially oriented template layers on nickel-based superalloy substrates" J. Mater. Res. **16**, 2175 (2001).

[4]S. Sathiraju, N.A. Yust, R.N. Nekkanti, I. Martense, A.L. Campbell, T.L. Peterson, T.J. Haugan, J.C. Tolliver, and P.N. Barnes, Mat. Res. Soc. Symp. Proc., **EXS-3**, pp. EE8.7.1-3 (2004).

[5]V.P.S Awana, A. Tulapukar, S.K. Malik, A.V. Narlikar, Role of Ca in enhancing the superconductivity of $YBa_2Cu_3O_{7-y}$" Phys. Rev. B **50**, 594 (1994)

[6]H. Hilgen kemp, C. W. Schneder, R. R. Shulz, B, Goetz A. Schemehl, H. Bielefeldt, J. Mannhart, "Possible solution of the grain-boundary problem for applications of high-T_c superconductors" Appl. Phys. Lett. **81**, 3209 (2002)

[7]G. Hammerl, A. Schmehl, R. R. Schulz, B. Goetz, H. Bielefeldt, C. W. Schneider, H. Hilgenkamp, J. Mannhart, "Enhanced super current density in polycrystalline YBa2Cu3O7-δ at 77 K from calcium doping of grain boundaries" Nature **407**, 162 (2000)

ELECTRODEPOSITED BIAXIALLY TEXTURED NI-W LAYER

Raghu Bhattacharya and Priscila Spagnol
National Renewable Energy Laboratory, 1617 Cole Boulevard, Golden, CO 80401

ABSTRACT
    Nonvacuum electrodeposition was used to prepare biaxially textured Ni-W coatings on Cu substrates. The samples were characterized by X-ray diffraction (including $\theta/2\theta$, pole figures, omega scans, and phi scans) and atomic force microscopy. Pole-figure scans show that electrodeposited Ni-W on textured Cu substrate is cube textured. Full-width at half-maximum values of the ω scan and ϕ scan of the electrodeposited Ni-W layers were comparable to the Cu base substrates, indicating good biaxial texturing.

INTRODUCTION

    It is now well established that biaxially textured crystalline substrates are critical to obtaining superior critical current densities (J_c) for $YBa_2Cu_3O_{7-\delta}$ (YBCO) superconductors. One way to accomplish biaxial texturing in a superconducting material is to grow epitaxial YBCO onto biaxially textured substrates. Rolling-assisted biaxially textured substrate (RABiTS) technology has proven to be very promising for fabricating YBCO-coated conductors[1-3] that can support large currents in high magnetic fields at 77 K. In this process, sharply biaxially textured Ni substrates are prepared by thermo-mechanical treatments; on which biaxially textured oxide buffer layers and YBCO films can be epitaxially grown. Recently, some attention has been given to Cu substrate coupled with a paramagnetic Ni-W layer instead of magnetic Ni to minimize the AC loss in the superconductor tape. At present, we are developing an electrodeposited Ni-W layer on Cu substrates. Electrodeposition is a potentially low-cost, nonvacuum, high-rate deposition process that can easily deposit uniform film on large nonplanar substrates. We already demonstrated a high-quality electrodeposited Ni layer on Ni-W that produced a critical current density of 1.8 MA/cm² at 75.2 K in a 600 gauss magnetic field for pulse-laser-deposited (PLD) $YBa_2Cu_3O_{7-\delta}/CeO_2/YSZ/CeO_2/ED-Ni/Ni-W$[4]. In this paper, we report on the electrodeposition of a Ni-W layer prepared on Cu substrates.

EXPERIMENTAL

    Electrodeposition (ED) of Ni was performed from a bath containing $NiSO_4 \cdot xH_2O$, $NiCl_2 \cdot 6H_2O$, and $Na_2WO_4 \cdot 2H_2O$ dissolved in deionized water. We prepared Ni-W films with varying concentration of Ni and W by adjusting the $Na_2WO_4 \cdot 2H_2O$ salt concentration in the plating bath solution. The films were deposited in a vertical cell in which the electrodes (both working and counter) were suspended vertically from the top of the cell. The ED precursors were prepared by a constant current mode in which the counter-electrode was a Pt gauze, and the substrate was Cu. An Arbin Instrument with a PC computer interface was used to prepare ED films. The ED deposition experiments were performed at 65°C and without stirring.

RESULTS AND DISCUSSIONS

    The thickness of the ED Ni-W films is about 1.0 μm. The optimized W concentration in the Ni-W film is about 9 atm%. The composition of the film is determined by inductively coupled plasma analysis. The as-electrodeposited Ni-W films are biaxially textured crystalline films. Figures 1a and 1b show the atomic force microscopy (AFM) images of the Cu substrate, and ED Ni-W on Cu substrate, respectively. The surface roughness of the Cu substrate and ED Ni-W are about 24 nm and 21.39 nm, respectively. The roughness of the electrodeposited layer was slightly improved compared to the Cu substrates. This slight roughness improvement could be due to auto chemical polishing of the substrate in the plating bath solution. It shows some outgrowth of ED Ni-W layers. We are resolving this problem by optimizing the electrodeposition condition.

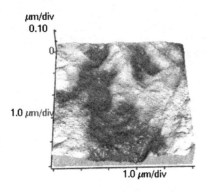

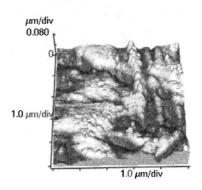

Figure 1a. AFM of Cu substrate.   **Figure 1b. AFM of ED Ni-W.**

The X-ray diffraction (XRD) scan of the Cu substrate and ED Ni-W on Cu are shown in Figs. 2a and 2b, and the pole-figure scans of Cu substrate and ED Ni-W are shown in Figs. 3a and 3b, respectively. The pole-figure scans show that as-deposited ED Ni-W on textured Cu substrate is biaxially textured. Full-width at half-maximum (FWHM) values of the ω scan (out-of-plane) and ϕ scan (in plane) of the Cu substrate were 9.9° (Fig. 4a) and 6.2° (Fig. 4b), and FWHM values of the ω scan (out-of-plane) and ϕ scan (in-plane) of the electrodeposited Ni-W layer were 9.4° (Fig. 5a) and 6.6° (Fig. 5b), respectively, which is comparable to the base Cu substrates and indicates good biaxial texturing.

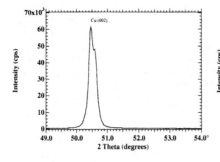

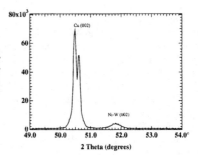

Figure 2a. XRD of Cu tape.   **Figure 2b. XRD of ED Ni-W.**

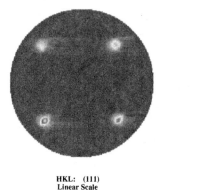
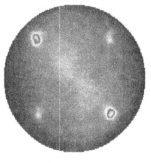

HKL: (111)
Linear Scale

HKL: (111)
Linear Scale

Figure 3a. Pole-figure scans of Cu substrate.

Figure 3b. Pole-figure scans of ED Ni-W.

CONCLUSION

We demonstrated that good-quality biaxially textured Ni-W could be prepared by potentially low-cost, nonvacuum electrodeposition. At present we are testing these electrodeposited layer by completing the buffer architecture and preparing a YBCO layer.

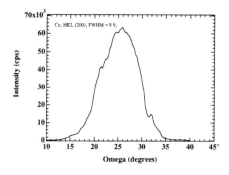
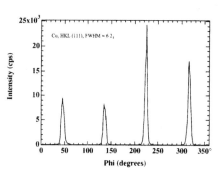

Figure 4a. Omega scan of Cu substrate.

Figure 4b. Phi scan of Cu substrate.

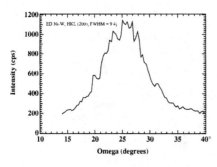

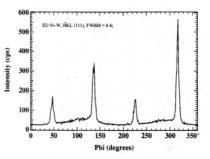

Figure 5a. Omega scan of ED Ni-W.   **Figure 5b. Phi scan of ED Ni-W.**

ACKNOWLEDGMENTS

This work was supported by the U.S. Department of Energy under Contract No. DE-AC36-99-GO10337. We like to thank Yust Nick and Paul Barnes (AFRL, Wright-Patterson AFB, OH) for providing textured Cu tape.

REFERENCES

[1] D.P. Norton, A. Goyal, J.D. Budai, D.K. Christen, D.M. Kroeger, E.D. Specht, Q. He, B. Saffian, M. Paranthaman, C.E. Klabunde, D.F. Lee, B.C. Sales, and F. List, *Science* **274** 755 (1996).

[2] J.E. Mathis, A. Goyal, D.F. Lee, F.A. List, M. Paranthanman, D.K. Christen, E.D. Specht, D.M. Kroeger, and P.M. Martin, *Jpn. J. Appl. Phys.* **37** L1379 (1998).

[3] C. Park, D.P Norton, D.T. Verebelyi, D.K. Christen, J.D. Budai, D.F. Lee, and A. Goyal, *Appl. Phys. Lett.* **76** 2427 (2000).

[4] R.N. Bhattacharya, J. Chen, P. Spagnol, and T. Chaudhuri, *J. Electrochem. Soc.*, to be published, 2004.

GROWTH OF Ba_2YNbO_6 BUFFER LAYERS BY PULSED LASER DEPOSITION ON BI-AXIALLY TEXTURED Ni ALLOY AND Cu- ALLOY SUBSTRATES

Srinivas Sathiraju,
National Research Council,
Propulsion Directorate
Air Force Research Laboratory
Wright-Patterson Air Force Base,
OH 45433, USA

Chakrapani V. Varanasi,
Univ. of Dayton Research Institute
Propulsion Directorate
Air Force Research Laboratory
Wright-Patterson Air Force Base,
OH 45433, USA

Nicholas A. Yust
Propulsion Directorate
Air Force Research Laboratory
Wright-Patterson Air Force Base
OH 45433, USA

Lyle B.Brunke,
University of Dayton Research Institute
Air Force Research Laboratory
Wright-Patterson Air Force Base,
OH-45433, USA

Paul N. Barnes
Propulsion Directorate
Air Force Research Laboratory
Wright-Patterson Air Force Base
OH-4433, USA

ABSTRACT

Epitaxial growth of Ba_2YNbO_6 (BYNO) thin films on biaxially textured Ni, Ni-W alloy and Cu-Fe alloy substrates is reported.. Commercially available Ni and Ni-W alloy substrates with a full width half maximum (FWHM) of 6 - 7° biaxial texture were used. A highly textured Cu-Fe alloy substrate with FWHM of 5 -7.5° was produced using the RABiTS process. BYNO films deposited on these Cu-Fe substrates have 8-10° FWHM texture. An attempt was made to deposit $YBa_2Cu_3O_{7-x}$ (Y-123) films on these buffer layers. Surface morphology of BYNO films and Y-123 films were investigated using a scanning electron microscope.

INTRODUCTION

The recent progress being made in the development of second generation coated conductor technology brings its incorporation into practical applications for the electrical power industry closer to reality. Recent advances on textured Ni and Cu alloys for the RABiTS (rolling-assisted biaxially textured substrate) architecture are showing promise as a potential substrate for the high temperature superconductor (HTS) coated conductors[1]. American Superconductor Inc. has demonstrated significant improvement of the overall performance using the RABiTs technique and has subsequently produced long lengths of YBCO coated conductors[2]. Among the several requirements for a practical coated conductor, a completely non-magnetic substrate tape for minimization of ferromagnetic losses is highly desirable in power applications where ac fields are present. The currently used Ni-W(5%) based tapes contribute to the overall total losses. Hence, Cu which is completely non-magnetic can not only eliminate the ferromagnetic losses of the substrate, but also has good thermal and electrical conductivity potentially serving as a stabilizing layer in addition to an epitaxial template. If so, the top stabilizing layer currently added to the coated conductor (Ag or Cu) can be eliminated. This will result in an increase in the engineering current if conducting buffer layers can be used between the superconductor and the metal substrate. Thus, the usage of a non-magnetic substrate such as Cu or Cu based alloys is encouraged for ac power applications[3,4]. Cantoni et al have used MgO as an oxygen barrier layer in buffer layer architectures intended for Cu[3]. Yang and Flynn [5] and Manning et al[6] have reported that the oxygen diffusion coefficient of MgO at 800 °C is nearly 13 orders of magnitude smaller than that of YSZ at the same temperature. Although MgO is known to grow on clean FCC metal surfaces, this layer alone is not a suitable sole buffer layer due to rapid Cu diffusion. For this reason, Cantoni et al have used TiN as the initial buffer layer between Cu and MgO[3]. Another possible solution is to use a different buffer layer which has less Cu diffusion at the given processing conditions. In this paper, we focused on using textured Ni-W and Cu-Fe alloys for depositing BYNO (transparent dielectric) buffer layers for the subsequent growth of Y-123 thin films. BYNO films are cubic and have a double perovskite structure (ABO_3) with a lattice parameter of ~ 0.84nm, which is double the size of the lattice parameter of MgO. Moreover, BYNO has a very stable phase up to 1200 °C and has excellent chemical compatibility with Y-123 thin films. In this paper, the growth optimization of BYNO buffer layers is reported using pulsed laser deposition on Ni, Ni-W, Cu, Cu-Fe textured substrates.

EXPERIMENTAL

In the present study, the films were prepared by pulsed laser deposition (PLD) using a Lambda Physic 301i excimer laser at the KrF, 248 nm, wavelength. In house prepared stoichiometric sintered YBNO targets were used. The BYNO targets were prepared by following the conventional solid state reaction route. High purity (99.99%) Y_2O_3, $BaCO_3$, and Nb_2O_5 powders were mixed in the stoichiometric ratio and then reacted at 1350° C for 48 hrs with three intermediate grindings. The phase purity of the calcified material was checked by x-ray diffraction. Phase pure material was finely grounded and pressed at a pressure of 3 Tons to form 1" diameter and 3mm thick discs. These discs were sintered again at 1400°C for 24 hrs. All the heat treatments were carried out in air. The Ni based substrates used in this study were obtained from Oxford instruments and Cu-based alloys were prepared in house[7]. The dimensions of the substrates used in this process were 10mm x 10mm x 0.25mm. The substrates were adhered to the substrate heater using silver paste. The base pressure of the chamber was approximately 10^{-7} Torr. Specimens were heated from room temperature to 800 °C in a 200 mTorr atmosphere of Ar+4% H_2 (forming gas) mixture to prevent oxidation of metal substrates.

The layers were applied in-situ in the following manner. After a soaking period of 10 min, the BYNO seed layer was deposited in the Ar+H_2 gas mixture for 5 minutes.

The chamber was then pumped down to 10^{-6} Torr and the deposition of BYNO was continued for an additional 5 min. The details of deposition parameters used are summarized in Table 1. After the deposition, films were cooled to room temperature in 500 Torr of oxygen. Y-123 films were deposited by PLD on these buffer layers using deposition conditions also given in Table 1.

The as deposited films were analyzed by x-ray diffraction techniques. Two-theta scans were accomplished using a Rigaku x-ray diffractometer. A Philips MRD with four circle diffractometer was used to study the crystalline alignment of the substrate, buffer layers, and the superconductor by means of omega, phi, and psi scans. The microstructure of the various films was evaluated under a scanning electron microscope (SEM) to study the surface roughness and surface morphology of the films grown in vacuum, in $Ar+H_2$, and in pure oxygen.

Table 1. Deposition conditions

Parameters	Ba2YNbO6	YBa2Cu3Oy
Deposition Temperature	RT – 850 °C	750-820 °C
Partial Pressure	200 mTorr	230mTorr
Laser Energy	2 J/Cm^2	2 J/Cm^2
Laser frequency	10 Hz	4-10Hz
Substrate-Target distance	6 cm	6 cm
Substrates	Rabits Ni, Cu, MgO	Buffered Ni substrates
Thickness	' 200 nm	200-500 nm

3. RESULTS AND DISCUSSION

a) Growth of BYNO films on Ni-based substrates:

X-ray diffraction of the films deposited at 700 °C was shown in figure 1. All the peaks were indexed based on a cubic perovskite structure with general formula $A_2BB'O_3$ with a lattice parameter of 8.4 Angstroms (PDF file no. 24-1042). Strong (400) and (220) orientations were observed at 29.9° and 42.9° of 2θ. However, weak reflections of 222 at 36.9° and another undesignated peak at 38.4° were also observed. The in plane orientations of BYNO are (400) (BYNO) // Ni (200).

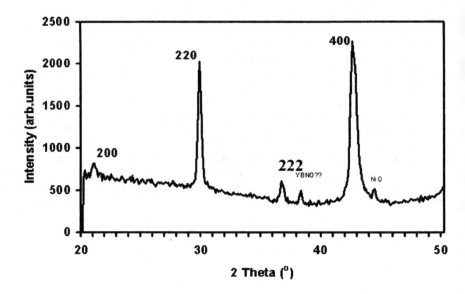

Figure1 X-ray diffractogram of BYNO thin film deposited on RABiTs Ni substrate deposited at 700 °C.

The (222) (12%) orientation of BYNO appear due to the surface energy and strain energy competition. At 700° C, low surface adatomic mobility may be responsible for the strong growth of (220) since this energy is not enough to overcome the surface diffusion activation energy. Hence, the film is less oriented in nature. BYNO films deposited around 800°C have shown highly (400) oriented films. Figure 2 shows the X-ray diffraction of the BYNO films deposited at 800 °C. It appears that the substrate temperature plays a critical role in the growth of the textured BYNO films. Y- 123 films were deposited on the substrates with BYNO buffer layer grown at 800 °C. To check the in-plane alignment of the films Φ (phi) scan measurements of (220) reflection of BYNO and (103) reflection of Y123 films was carried out. The FWHM of BYNO films is around 12° and of Y-123 films is around 10°. The best texture obtained by BYNO films is 8° using IBAD substrates.[8] Figure 3 shows the phi scan of (103) reflections of Y-123 films deposited on BYNO buffered RABiTS Ni. However, the superconducting quality of these films was found to be poor. Process optimization is presently being done. Figure 4 shows the SEM of the BYNO film surface morphology deposited on Ni substrate in Ar+H_2 medium.

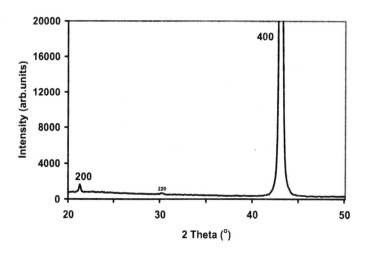

Figure 2. XRD of BYNO film deposited at 800 °C on Ni substrate

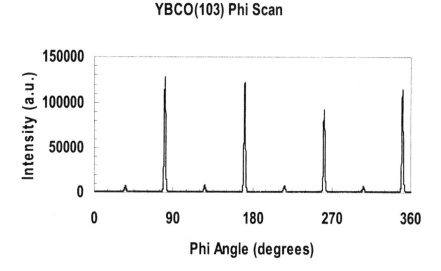

Figure 3. Φ Scan of (*103*) reflection of Y-123 film deposited on BYNO buffered Ni substrate

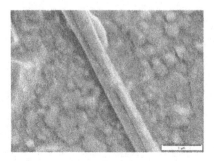

Figure 4 Surface morphology of BYNO film deposited on RABiTs Ni

b) Growth of BYNO films on Copper substrates

X-ray diffraction of the BYNO films deposited on textured Cu substrates at 680 °C was shown in figure 5. Strong (*220*) and (*400*) reflections at 30° and 43° respectively were observed. Weak reflections of (*222*) and Cu peak are observed at 36.8° and 44.8° respectively.

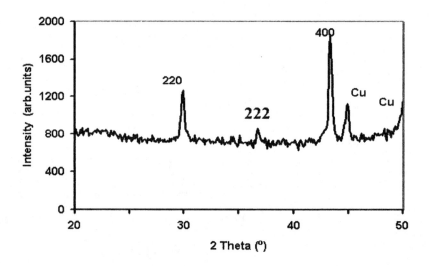

Figure 5.   X-ray diffraction pattern of BYNO film deposited at 680 °C on RABiTS Cu substrate.

When the substrate temperature is around 780°C, highly (*400*) oriented growth of the films was observed. Figure 6 shows the X-ray diffraction of the BYNO films deposited at 780 °C. In this case also, we do have (*220*) orientations. But when compared with the intensity of (*400*) orientation (*220*) peak intensity is almost negligible.

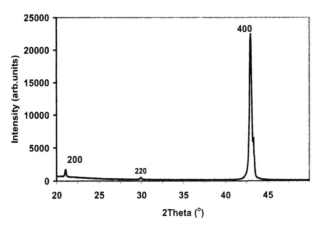

Figure6. **X-ray diffraction pattern of BYNO film deposited at 780 °C on Cu
substrate**

A SEM micrograph of BYNO film deposited on Cu-Fe substrate in Ar+H₂
medium is shown in Figure 7. The surface morphology indicates a relatively smooth
surface with very minimal particulate growths on the surface. Figure 8 shows Y-123 film
deposited in the presence of oxygen on the BYNO buffer layer. It was observed that in
films the adhesion of Y-123 films is very poor and as a result several peeled off regions
were observed. Efforts are underway to grow Y-123 films on these BYNO buffer layers.

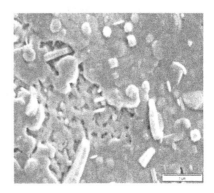

Figure 7. SEM micrograph of BYNO film
deposited at 800 °C on RABiTS Cu-Fe
substrate.

Figure 8. SEM Picture of Y-123
film deposited on BYNO buffered
RABiTs Cu-Fe substrate.

CONCLUSIONS

In conclusion, we report the growth conditions and microstructural properties of BYNO films deposited on RABiTs Ni and Cu alloy substrates. Films deposited around 800 °C substrate temperature have shown highly (400) oriented BYNO films. The FWHM of these films from phi scans on both substrates is around 12°. The best microstructure of BYNO films was observed when films were grown in the Ar+H$_2$ reduced atmosphere using pulsed laser deposition. Further growth optimization of BYNO buffers and Y-123 films on these buffer layers is in progress.

ACKNOWLEDGMENTS

Srinivas acknowledges National Research Council (NRC), National Academy of Sciences (NAS) and Propulsion Directorate, Air Force Research Laboratory for the Senior Research Associate fellowship and coated conductor research support.

REFERENCES

[1] A. Goyal, D.P.Norton, J.D. Budai, M. Paranthaman, E.D. Specht, D.M. Kroeger, D.K. Christen, Q. He, B. Saffian, F. A. List, D.F. Lee, P.M. Martin, C. E. Klabunde, E. Hartfield, and V. K. Sikka "High critical current density superconducting tapes by epitaxial deposition of YBa$_2$Cu$_3$O$_x$ thick films on biaxially textured metals Appl. Phys. Lett. **69,** 1795 (1996)

[2] M.W. Rupich, Q.Li, S. Annavarapu, C. Thieme, W. Zhang, V. Prunier, M. Paranthaman, A. Goyal, D.F. Lee, E.D. Specht, and F.A. List, "Low cost Y-Ba-Cu-O coated conductors" IEEE trans. on Appl. Supercon. **11,** 2927 (2001)

[3] C. Cantoni, D.K. Christen, M. Varela, J.R. Thompson, S.J. Pennycook E.D.Specht, A. Goyal Deposition and characterization of YBa$_2$Cu$_3$O$_{78}$/ LaMnO$_3$/ MgO/TiN heterostructures on Cu metal substrates for development of coated conductors, J. Mater. Res. **18** 2387, (2003).

[4] S. Pinol, J. Díaz, M. Segarra, F. Espiell, "Preparation of biaxially cube textured Cu substrate tapes for HTS coated conductor applications", Super Cond. Sci. and Tech. **14**(2001)11.

[5] M.H. Yang and C.P. Flynn, "Ca^{2+} and ^{18}O^{2-} diffusion in ultrapure MgO", J. Phys. Condens. Matter, **8,** L279, 1996.

[6] P.S. Manning, J.D. Sirman, and J. A. Kilner, "Oxygen self-diffusion and surface exchange studies of oxide electrolytes having the fluorite structure" Solid State Ionics**, 93**, 125, (1997).

[7] S.Sathiraju, N. Yust, R. Nekkanti, I. Maartense, A.Campbell, T. Peterson, T. Haugan, J.Tolliver, P. Barnes, Growth and Microstructural studies of YBa$_2$NbO$_6$ thin films" Mat. Res. Soc. Symp. Proc., **EXS-3**, pp. EE5.6. 21 (2004).

[8] N.A. Yust, R. Nekkanti, L.B. Brunke, R. Srinivasan, and P.N. Barnes, *Copper Metallic Substrates for HTS Coated Conductors*, submitted; also N. Yust, R. Nekkanti, L. Brunke, and P. Barnes, Mat. Res. Soc. Symp. Proc., **EXS-3**, pp. EE8.7.1-3 (2004).

Bulk Superconductors

COARSENING OF $BaCeO_3$ AND Y_2BaCuO_5 PARTICLES IN $YBa_2Cu_3O_{7-x}$ SEMISOLID MELT

Oratai Jongprateep
Department of Ceramic Engineering
University of Missouri-Rolla
Rolla, MO 65401

Fatih Dogan
Department of Ceramic Engineering
University of Missouri-Rolla
Rolla, MO 65401

ABSTRACT

In an attempt to enhance the critical current density in melt textured $YBa_2Cu_3O_{7-x}$ (Y123), coarsening behavior of non-superconducting inclusions in the semisolid melt of $YBa_2Cu_3O_{7-x}$ was investigated. Coarsening behavior of $BaCeO_3$ was compared with that of Y_2BaCuO_5 (Y211) particles above the peritectic decomposition temperature as a function of time. Microstructural characterization of the quenched samples revealed a significant coarsening of Y211 phase, while the size of $BaCeO_3$ particles remained in the sub-micrometer range.

INTRODUCTION

High critical current densities of high temperature superconductors are necessary for various practical applications such as frictionless bearings used for development of flywheel energy storage systems. It is well known that critical the current density (J_c) of Y123 is enhanced by the presence of effective flux pinning sides such as non-superconducting particles, twin boundaries and defect structure [1-2]. The interface between the superconducting matrix and small-sized, uniformly distributed second phase particles can be very effective for magnetic flux pinning and lead to enhanced superconducting properties.

Doping of Y123 with cerium has been known to improve the critical current density in Y123 [3]. It is believed that cerium-based compounds act as growth inhibitors of the Y211 particles, thus enhancing the refinement of Y211 particles [4-7]. Formation of sub-micron sized $BaCeO_3$ particles in Y123 results in further increase of flux pinning centers.

Relationship between the microstructural development and superconducting properties of cerium doped Y123 has been reported previously [8-10]. However, the effect of the processing parameters on the liquid loss and microstructural development of melt textured Y123 doped with cerium oxide have not been well established. This study addresses coarsening of $BaCeO_3$ and Y211 particles in the semisolid melt as a function of the dopant concentration and the holding time.

EXPERIMENTAL PROCEDURE

The Y123 doped with $BaCeO_3$ powder was prepared by mixing commercially available Y_2O_3, CuO, $BaCO_3$ and CeO_2. The starting compositions were Y123+0.5 wt% $BaCeO_3$, Y123+5 wt% $BaCeO_3$ and Y123+10 wt% $BaCeO_3$. The powders were uniaxially pressed to obtain pellets with 10 mm in diameter.

The powder compacts were heated from room temperature to $1050°C$ at a heating rate of $47°C/hr$. Each set of samples was subsequently held at $1050°C$ for 0.5, 48 and 96 hours. The samples were cooled rapidly to $1010°C$ and subsequently air quenched to the room temperature. The microstructure of quenched samples was characterized by scanning electron microscopy (SEM JEOL-T330A) techniques. The size distribution of Y211 and $BaCeO_3$ particles were analyzed using Scion Image Software. Due to the highly anisotropic shape of needle-like Y211 phase, the width of the particles was measured for the analysis. A minimum of 100 particles were counted to obtain the size distribution of BaCeO$_3$ and Y211 particles.

RESULTS AND DISCUSSION

Effect of cerium oxide doping on Y211 size and morphology:

Figure 1 shows the effect of cerium oxide addition on the size and morphology of needle-like Y211 particles (particle width) in melt quenched samples held at 1050°C for 0.5 hour. Increasing amount of the dopant addition results in the formation of BaCeO$_3$ particles as fine and faceted particles.

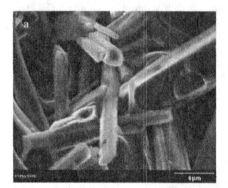

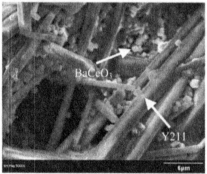

Figure 1. (a) (b) and (c). 211 size and morphology of melt quenched samples with 0.5 wt%, 5 wt% and 10 wt% BaCeO$_3$, respectively.

The size of Y211 particles, as shown in Figure 2, is within the range of 1.3 μm–2.2 μm. While the average Y211 particle size is approximately 2 μm with the lowest amount of BaCeO$_3$ doping (0.5 wt%), higher dopant concentrations (5% and 10%) led to further refinement of Y211 particles.

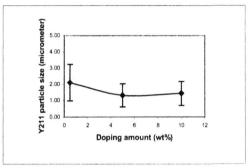

Figure 2. Effect of BaCeO₃ doping amount on average size of Y211 (particle width) for melt quenched samples held at 1050°C for 0.5 hr

Effect of Holding Time on the Size Distribution of Y211 Particles:

Figure 3 shows the microstructural development of melt quenched samples with 10 wt% BaCeO₃ addition at different holding times. Due to the significant differences in the size of Y211 particles, these SEM micrographs were taken at different magnifications. Longer holding times resulted in significant coarsening of Y211 phase.

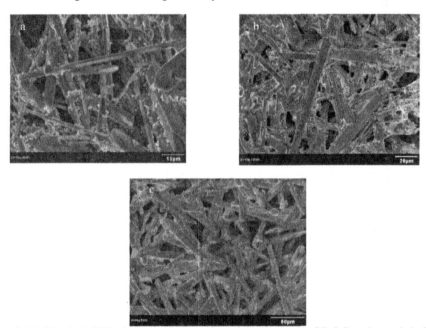

Figure 3. (a) (b) and (c). Y211 size and morphology of samples with 10 wt% of BaCeO₃ melt quenched after holding time for 0.5, 48 and 96 hrs, respectively.

The effect of the holding time on the size of Y211 particles in melt quenched samples is shown in Figure 4. The size of Y211 particles increases from approximately 2.1 to 9.3 μm as the holding time in the melt increases.

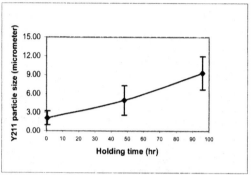

Figure 4. Effect of holding time on average particle size of Y211 for samples doped with 10 wt% BaCeO₃

Particle size distribution of Y211 after a holding time of 0.5, 48 and 96 hours at 1050°C are shown in Figure 5. The size distribution of Y211 particles is more uniform after a short holding time, while longer holding times results in a broader size distribution.

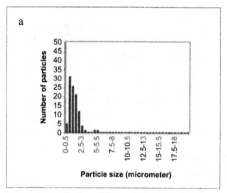
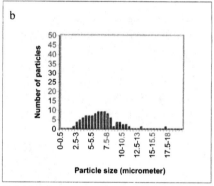

Figure 5. Distribution of 211 particle width of the sample with holding time (a) 0.5 hr and (b) 96 hr, respectively

Effect of Holding Time on the Size and Morphology BaCeO₃:

Figure 6a-c shows that the size and morphology of BaCeO₃ particles in the melt quenched samples. As the holding time in the melt increases from 0.5 hour to 96 hours, the size of BaCeO₃ particles with a cubic morphology increases only slightly but remains mainly in the sub-micrometer range.

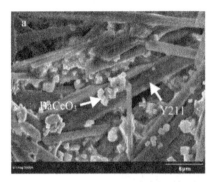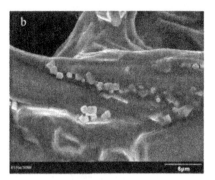

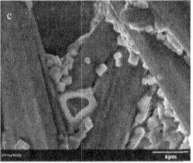

Figure 6. (a) (b) and (c). SEM micrographs shows $BaCeO_3$ particles in samples with 10 wt% of $BaCeO_3$ prepared at holding time of 0.5, 48 and 96 hr respectively

Figure 7 and 8 show the effect of holding time on the particle size distribution of $BaCeO_3$ phase. While the average size of $BaCeO_3$ particles with a narrow distribution is approximately 0.6 µm after a holding time for 0.5 hours, the average particle size increases slightly to 0.9 µm with a broader distribution after 96 hours.

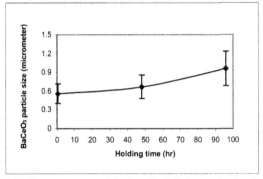

Figure 7. Effect of holding time on average particle size of $BaCeO_3$ for samples doped with 10 wt% $BaCeO_3$

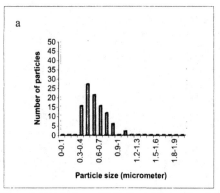 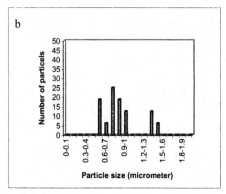

Figure 8. Distribution of BaCeO$_3$ particle width of the sample with holding time (a) 0.5 hr and (b) 96 hrs, respectively

To obtain higher critical current densities in Y123 superconductors, nanometer sized non-superconducting inclusions can play an important role as effective flux pinning sites. The microstructure of melt quenched Y123 samples with BaCeO$_3$ addition reveals refinement of Y211 particles in the presence of cerium doping. Cerium-based compounds can alter the kinetics of the Y211 coarsening, according to the equation 1:

$$R = (\sigma/\eta)^{(n)} \qquad\qquad (1)$$

Where R is the mean radius for Y211 particles, σ is the interfacial energy between the Y211 surface and liquid phase, η is the viscosity of the liquid phase and n is a constant [4]. It is believed that addition of cerium-based compounds can reduce the interfacial energy, while increasing the melt viscosity. Thus, the coarsening of the Y211 particle is hindered as a result of BaCeO$_3$ addition [4].

While Y211 particles exhibit a large coarsening effect as the holding time increases, the size of BaCeO$_3$ particles remains mostly in the sub-micrometer range as the holding time increases. From the observation above, it can be inferred that fine BaCeO$_3$ inclusions can act as effective flux pinning centers towards improved superconductor properties in Y123. Further studies are in progress to grow melt textured single crystals of Y123 doped with BaCeO$_3$.

SUMMARY

The relationship between the amount of BaCeO$_3$, holding time and microstructural development of melt quenched Y123 was investigated. It was shown that cerium addition to Y123 reduces the coarsening of Y123 particles in the semisolid melt during short holding times. While the long holding times in the melt leads to significant coarsening of Y211 particles, the size of BaCeO$_3$ inclusions remain in sub-micrometer range which could become effective flux pinning centers in melt textured Y123 superconductors.

ACKNOWLEDGEMENTS

This work was supported by Boeing Phantom Works and the Materials Research Center at the University of Missouri-Rolla.

REFERENCES

[1] M. Murakami, "Processing of bulk YBaCuO," *Supercond. Sci. Technol.*, **5**, 185-203 (1992).

[2] A. M. Campbell, J. E. Evetts, and D. Dew-Hughes, "Pinning of flux vortices in Type II superconductors" *Phil. Mag.*, **18**, 313 (1968).

[3] S. Pinol, F. Sandiumenge, B. Martinez, V. Gomis, J. Fontcuberta, X. Obradors, E. Snoeck and C. Roucau, "Enhanced Critical Current by CeO_2 Additions in Directionally Solidified $YBa_2Cu_3O_7$," *Appl. Phys. Lett.*, **65** [11], 1448-50 (1994).

[4] S. Marinel, I. Monot, J. Provost and G. Desgardin, "Effect of SnO_2 and CeO_2 Doping on Microstructure and Superconducting properties of Melt Textured Zone $YBa_2Cu_3O_{7-x}$," *Supercond. Sci. Technol.*, **11**, 563-72 (1998).

[5] C. J. Kim, K. B. Kim and G. W. Hong, "Y_2BaCuO_5 Morphology in Melt-Textured Y-Ba-Cu-O Oxides with $PtO_2.H_2O/CeO_2$ Additions," *Physica C*, **232**, 163-72 (1994).

[6] S. Pinol, F. Sandiumenge, B. Martinez, N. Vilalta, X. Granados, V. Gomis, F. Galante, J. Fontcuberta and X. Obradors, "Modified Growth Mechanism in Directionally Solidified $YBa_2Cu_3O_7$," *IEEE Trans. Appl. Supercond.*, **5**, 1549-52 (1995).

[7] J. Reding and F. Dogan, "Microstructural Evolution of Melt Textured $YBa_2Cu_3O_{7-x}$ Doped with $BaCeO_3$," *Ceram. Trans.*, **112**, 509-15 (2000).

[8] C. Harnois, M. Hervieu, I. Monot and G. Desgardin, "Relations between Microstructure and Superconducting Properties of Ce+Sn Doped YBCO," *J. Euro. Ceram. Soc.*, **21**, 1909-12 (2001).

[9] C. J. Kim and P.J. McGinn, "Anomalous Magnetization and Oxygen Diffusion in the Melt-Textured Y-Ba-Cu-O Domain with/without $BaCeO_3$ Addition," *Physica C*, **222**, 177-83 (1994).

[10] M P Delamare, I Monot, J Wang, J Provost and G. Desgardin, "Influence of CeO_2, $BaCeO_3$ or PtO_2 Additions on the Microstructure and the Critical Current Density of Melt Processed YBCO Samples," *Supercond. Sci. Technol.*, **9**, 534-42 (1996).

THE MICROSTRUCTURE AND SUPERCONDUCTING PROPERTIES OF $YBa_2Cu_3O_y$ - BASED CERAMICS

E.D. Politova, G.M. Kaleva, K.I. Furaleva and S.G. Prutchenko
L.Ya.Karpov Institute of Physical Chemistry
10 Vorontsovo pole st.
Moscow, 105064, Russia

ABSTRACT

The influence of various additives and aliovalent ions substitution on processing, structure parameters, microstructure and superconducting properties of $YBa_2Cu_3O_{7-\delta}$ (YBCO) ceramics has been studied. The stabilizing effect of the CuO additive on the microstructure and superconducting properties has been revealed. Fluorine introduction has been shown to deteriorate superconducting properties, though the existence of the two-dimensional species with higher superconducting temperatures has been confirmed. The microstructure regulation has been also proved to improve the transport current ability of polycrystalline ceramics.

INTRODUCTION

To promote industrial applications, the relationship between the density of the critical current J_c and the composition, processing and microstructural features of polycrystalline $YBa_2Cu_3O_{7-\delta}$ (YBCO) ceramics should be understood[1]. The control of both the composition and preparation routes is known to be the key factor for the regulation of the flux pinning in the superconducting ceramics.

The formation of additional pinning centres by the introduction of submicron size additives has been reported[2-5]. The presence of submicron inclusions of the dielectric Y_2BaCuO_5 phase which is a product of peritectic decomposition of the YBCO phase has been shown to be responsible for the increasing current ability[6]. The size regulation of the Y_2BaCuO_5 inclusions is possible if special microadditives are used[3]. Variations of the melt-growth method allow the optimization of textured ceramics with "clean" itergrain boundaries[7-9].

In this work, ceramic solid solutions $YBa_2Cu_{3+y}O_{7-\delta}$ with $-0.05 \leq y \leq 0.3$ (I); $YBa_2Cu_3O_{7-\delta}F_z$ with $z<0.4$ (II); $(1-x) YBa_2Cu_3O_{7-\delta} - xBaTiO_3$ with $x<0.03$ (III); $(1-x) YBa_2Cu_3O_{7-\delta} -x ScSr_2V_3O_{11}$ with $x \leq 0.10$ (IV) have been prepared. The influence of the copper excess, the fluorine introduction in the lattice, as well as the highly dispersed $BaTiO_3$ and complex $ScSr_2V_3O_{11}$ additive on the superconducting temperature, microstructure and transport properties of YBCO ceramics has been studied.

EXPERIMENTAL

The modified ceramics I-IV have been prepared by solid state reactions using initial mixtures of oxides and carbonates. To prepare the fluorine containing ceramics, appropriate quantities of barium fluorine acetate $Ba(CF_3COO)_2$ (BFA) substituted for $BaCO_3$ in initial mixtures to obtain compositions $YBa_2Cu_3O_{7-\delta}(F_z)$ $(z<0.4)$. The BFA decomposes at temperatures higher than 400°C with formation of extremely reactive BaF_2. The highly dispersed $BaTiO_3$ used has been prepared from alkoxide precursors[11].

The phase content and orthorhombic unit cell parameters were controlled by the X-ray diffraction method (Cu-K_α beam); microstructure and the composition of individual grains checked by the scanning electron microscopy method (JEOL-35CF equipped with the LINK set).

The superconducting temperature T_c has been measured using both the four probe dc- and ac-current (f=1 MHz) methods. The critical current density of the ring shaped samples has been measured by the contactless transformer method[12].

RESULTS AND DISCUSSION

The influence of CuO excess on microstructure and superconducting properties of $YBa_2Cu_{3+y}O_{7-\delta}$ ceramics

The typical X-ray diffraction pattern for the orthorhombic perovskite YBCO ceramics is shown in Fig. 1. The ceramics prepared with the CuO deficiency revealed degradation of superconducting (SC) properties. In compositions with x<-0.003 the specific resistance value at room temperature $\rho_{r.t.}$ increases, the metallic type of $\rho(T)$ behaviour changes to a semiconductive one, T_c value decreases, and SC transition diffuses due to the increased concentration of defects in Cu sites and the increased amount of dielectric impurity phases as well[13-15].

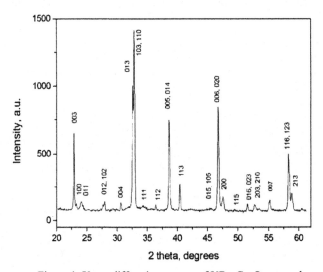

Figure 1. X-ray diffraction pattern of $YBa_2Cu_3O_{7-\delta}$ sample.

A slight CuO excess provides the optimal oxygen content. This is testified by the parameter c decreasing and the b/a ratio increasing (Fig. 2). For such ceramics, low $\rho_{r.t.}$ values, rather sharp SC transition and the increasing share of SC phase estimated via the ac-current inductance method (Fig. 3). The T_c increasing observed in the case of the overstoichiometric CuO addition may be primarily explained by the filling the Cu vacant sites which usually occur in YBCO samples prepared from the stoichiometric mixtures. It is worth noting that the introduction of large CuO amounts (x>0.1) facilitates formation of the local eutectiques $BaCuO_2$-CuO,

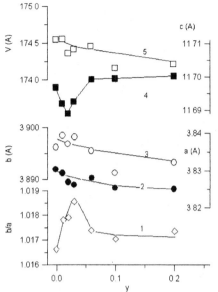

Figure 2. The unit cell parameters b/a (1), a (2), b (3), c (4), V (5) versus the CuO content y plot for YBa$_2$Cu$_{3+y}$O$_{7-\delta}$ ceramics.

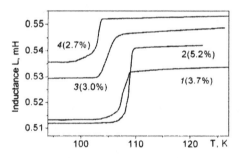

Figure 3. The temperature dependences of the ac-inductance for ceramics YBa$_2$Cu$_{3+y}$O$_{7-\delta}$ with y=0.06 (1); (1-x) YBa$_2$Cu$_3$O$_{7-\delta}$ - x BaTiO$_3$ with x=0.005 (2); YBa$_2$Cu$_3$O$_{7-x}$F$_x$ with x=0.04 (3); 0.1 (4). Figures in brackets define the normalized $\Delta L/L$ value.

YBa$_2$Cu$_3$O$_y$ - CuO, YBa$_2$Cu$_3$O$_y$ - BaCuO$_2$ with low melting temperatures that results in an uncontrolled increase in size of separate grains (Fig. 4).

The influence of fluorine on super conducting properties
It has been shown that the fluorine content substituting for oxygen in the YBCO lattice does not exceed ~3 at. %. This value corresponds to x=0.1 in the initial mixtures. The unit cell parameters remain constant in the pure perovskite phase samples with x<0.08. However, in compositions with x>0.08 a sharp decrease in the b/a ratio along with an increase in the parameter c and the unit cell volume are observed (Fig. 5). A T_c decrease with increasing x up to 0.1 correlates with the unit cell parameters changes and evidently indicate the fluorine filling for the oxygen vacant positions O5 in the basal plane. In turn, this determines deterioration of SC properties. However, the T_c value remains higher than 77 K in multiphase samples with 0.1<x<0.4. All the samples with x<0.4 reveal typical metallic $\rho(T)$ behavior. The increasing $\rho_{r.t.}$ value indicates the decreasing volume share of the SC phase (Fig. 3).

Surprisingly, high T_c value ~105-115K has been observed in some multiphase compositions with x=0.14-0.15 when they were tested within ~36 hours after the thermal treatment (Fig. 6). As it can be seen from the curves 1, 2 and 4 in Fig. 6, the zero resistance is reached at 105

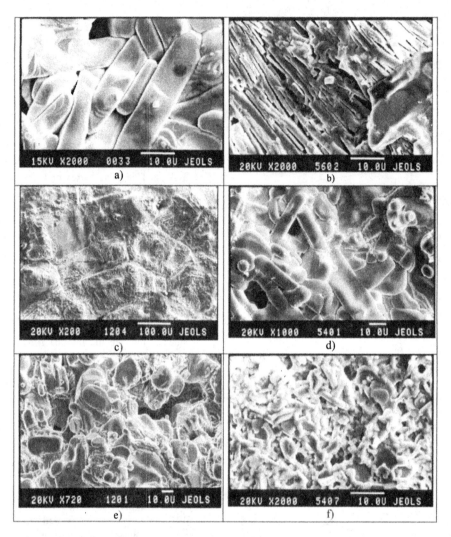

Figure 4. Scanning electron micrographs of as fired surfaces for ceramics $YBa_2Cu_{3+y}O_{7-\delta}$ with $x=0.1$ (a), 0.06 (b), $(1-x)$ $YBa_2Cu_3O_{7-\delta} - x$ $BaTiO_3$ with $x=0.01$ (c), $YBa_2Cu_3O_{7-\delta}F_z$ with $z=0.1$ (d), (e), $(1-x)$ $YBa_2Cu_3O_{7-\delta} - x$ $ScSr_2V_3O_{11}$ with $x=0.15$ (f).

and 115 K respectively. These curves are rather smooth and do not reveal the part with the sharp $\rho(T)$ decreasing. However, the $\rho(T)$ dependences measured for these samples after ~130 h after sintering (curve 3 in Fig. 6) reveal the normal metallic type $\rho(T)$ behavior with $\rho_{300K}/\rho_{100K}=2$, and conventional SC transition with $T_c=89$ K and $\Delta T=4$ K. The effect observed has been explained by

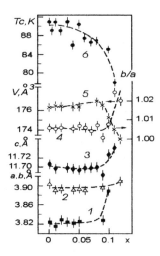

Figure 5. Dependences of the unit cell parameters a (1), b 2), c (3) V (4), b/a (5) and T_c (6) versus nominal fluorine content z for $YBa_2Cu_3O_{7-z}F_z$.

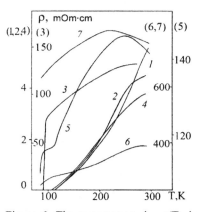

Figure 6. The temperature dc- $\rho(T)$ dependences for $YBa_2Cu_3O_{7-z}F_z$ ceramics with z=0.15 (1-4), 0.11 (5), 0.14 (6), 0.16 (7) measured in 35 h (1-7) or 130 h (3) after the thermal treatment.

a modification (melting or amorphization) of the grain surface in the presence of fluorine species. As other possible reasons for the effect observed, coherent phenomena on the interphase boundaries between SC grains and the ionic conductor BaF_2 grains or the creation of low dimensional "short living" SC phases with high T_c inside the grains may be examined as well.

Though the volume share of such grains is rather small and could not be detected by the Meissner effect, the phases revealed are reproducible. Apparently, similar phases were observed by Ovshinskii[16]. It allows us to consider the oxides studied as substances in which superconducting two dimensional phases may occur on the modified surface of the grains. This conclusion is supported by the theoretical results on the possibility of the superconducting transition realization in the case of the characteristic surface states creation which occur in a two dimensional conductor under certain conditions[17].

The influence of complex additives on microstructure and SC properties

Effects of various additives and sintering conditions on microstructure parameters, T_c and J_c values have been studied for the systems I-IV [11,14,15,128,19]. The choice of these systems is mostly motivated by the reported effects of the variation in the microstructure when the sample composition is changed. As can be seen in Fig. 4, the average grain size is changed drastically. Moreover, using the additional partial melting of ceramics by heating them to temperatures higher than peritectic decomposition temperatures, up to ~1020-1040°C, the ceramics with highly textured surface could be prepared (Fig. 4, b).

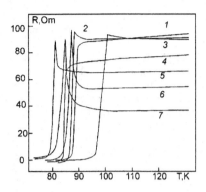

Figure 7. Temperature dependences of the active losses $R(T)$ measured by the ac-inductance technique for ceramics $(1-x)$ $YBa_2Cu_3O_{7-\delta}$ - x $ScSr_2V_3O_{11}$ with $x=0$ (1,4), 0.04 (2), 0.06 (3), 0.10 (5), 0.08 (6) prepared at $t_2=930$ (1,6), 950°C (2-5) and for ceramics $Bi_2Sr_2CaCu_2O_y$ (7).

According to the X-ray diffraction and the electron microprobe data, the substitution of Ti for Cu and Sr for Ba takes place in the samples III and IV respectively. The samples III are characterized by the enlargement of grains, while a decrease in the grain size practically by an order of magni-tude is observed for the samples IV when x increases to 0.15 (Fig. 4 c, f). The microstructure changes observed in the system IV are governed by the inhibition effect characteristic for modi-fied YBCO ceramics in the presence of the admixture phases. The growth of grains in the samples is conditioned by the increasing number of local eutec-tiques CuO-$YBa_2Cu_3O_y$ or by the change in the peritectic decomposition temperature in solid solutions on the basis of the YBCO phase. The texture

observed on the surface of the samples studied arises due to the liquid phase presence as well. Substitution of strontium for barium is also responsible for the grain growth inhibition effect in the system IV.

The critical current density j_c for the ring shape samples has been measured by the ac-contactless transformer method. Dependence of j_c on the thickness of the ring d has been confirmed. It is described by the function:

$$j_c/j_{co} \sim d^{1/3} \tag{1}$$

This dependence is conditioned by the influence of the self magnetic field of the current flowing through the sample and is characteristic of the critical state model in which the local current density is connected with the magnetic field susceptibility B as:

$$j_c(B)=j_o/[1+(B/B_o)^2] \tag{2}$$

The observed variations in grain sizes resulted in appearance of peaks near the T_c on the temperature dependences of the active losses curves measured by the ac- inductance method (Fig. 7). It has been confirmed that the temperature dependence of the real part of impedance has maximum near the T_c[20, 21]. Maximum in the electromagnetic field energy dissipation takes place when the depth of the electromagnetic field penetraton into the intergrain region of a sample λ^T becomes equal to the half thickness of the sample $d/2$. The maximum in the active losses occurs also in the case when the main part of dissipation losses is determined by the movement of the magnetic flux inside the grains. In this case the penetration depth into the grain λ_g becomes equal to the average grain radius $<D>$. It has been also revealed that at measuring frequencies higher than 1 MHz, the maximum in the real part of the impedance versus temperature curve corres-ponded to entrapping the magnetic field by the grains of ceramics[22].

It is worth noting that the maximum in the active losses curves has not been observed for large grain ceramics while it is a characteristic feature of the 2212 Bi-containing ceramics with specific plate-like grains (curve 7 in Fig. 7). The increasing maximum weight indicates the increasing penetration depth λ_g, that is, the change in the characteristic parameter $\lambda_g/<D>$ value that determines the real and imaginary parts of the magnetic susceptibility of the ceramics.

The highest values of the transport current $j_c>1\cdot10^3$ A/sm^2 have been observed both for the dense ceramics I with the specific plate-like microstructure (Fig. 4 b) and for ceramics with a small grain size (Fig. 4 f). According to the electron microprobe analysis yttrium concentration increasing by more than 10 % has been revealed in the grains of ceramics III which are character-ized by high j_c values. As the yttrium concentration practically close to the stoichiometric value equal to 1 was observed in the large grains ceramics, we believe this effect to be indicative of the creation of additional dielectric pining centers for magnetic field in the fine grain ceramics.

Finally, the results obtained support the conclusion that high j_c values of bulk YBCO ceram-ics are ensured if the controlled microstructure and density of ceramics are provided by varying the concentration of some additives and the thermal annealing process.

REFERENCES

[1]D.Larbalestier, A.Gurevich, D.Matthew, A.Polynskii, High-Tc superconducting materials for electric power applications, *Nature*, **414** [11] 368-377 (2001).

[2]T. Meignan, A.Banerjee, J.Fultz, P.J.McGinn, Effects of Ce-based additions during texturing of YBa$_2$Cu$_3$O$_{7-\delta}$, *Physica C*, **281** [2-3] 109-20 (1997).

[3]C.-J.Kim, K.-B.Kim, I.-H.Kuk, G.-W.Hong, Role of PtO$_2$ on the refinement of Y$_2$BaCuO$_5$ second phase particles in melt-textured Y-Ba-Cu-O oxides, *Physica C*, **281** [2-3] 241-52 (1997).

[4] Q. Chen, M.Fang, Z.Liao *et al.*, Influence of nanometersized MgO particles on flux pinning in YBa$_2$Cu$_3$O$_{7-\delta}$, *Physica C*, **227** [1-2] 113-118 (1997).

[5]T.Hamada, Y.Ohzono, T.Akune, N.Sakamoto, Effect of Platinum on the critical current den-sity of fluorine-doped YBa$_2$Cu$_3$O$_x$ superconductors, *J.Mater.Sci.* **32** [13] 3469-73 (1997).

[6]T.Aswlage, K.Keefer, Liquidus relations in Y-Ba-Cu oxides, *J.Mater.Res.* **3** [6] 1279-91 (1988).

[7]A.Gencer, A.Ates, E.Aksu et.al., Microstructural and physical properties of YBa$_2$Cu$_3$O$_{7-\delta}$ su-perconductors prepared by the Flame-Quench-Melt-Growth (FGMG) method, *Physica C*, **279** [3-4] 165-72 (1997).

[8]O.F.Schilling, Y.Yang, C.R.M.Grovenor, C.Beduz, Characterization and properties of melt grown Y-Ba-Cu-O samples containing titanate inclusions, *Physica C*, **170** [1/2] 123-129 (1990).

[9]S.Lin, T.H.Tiefel, R.C.Sherwood et al., Fabrication of dense Ba$_2$YCu$_3$O$_{7-\delta}$ superconductor wire by molten oxide processing, *Appl.Phys.Lett.*, **51** [12] 943-45 (1987).

[10]G.M.Kaleva, S.G.Prutchenko, T.A.Starostina, B.Sh.Galyamov, E.D.Politova, Phase content, structure and superconducting propertuies of the fluorine-substituted yttrium-barium cuprate, *Russian Journal of Inorganic Materials*, **28** [7] 1426-1430 (1992).

[11]G.M.Kaleva, K.I.Furaleva, S.G.Prutchenko, E.D.Politova, Study of influence of high dis-

perse additives $BaTiO_3$ and TiO_2 on microstructure and superconducting current density of the $YBa_2Cu_3O_{7-\delta}$ ceramic, *Russian Journal of Inorganic Materials*, **37** [1], 71-76 (2001).

[12]G.P.Meisner, C.A.Taylor, Factors affecting critical currents determined by the transformer method for the $YBa_2Cu_3O_y$ high-Tc superconductors. *Physica C,* **169** [5-6] 303-13 (1990).

[13]E.D.Politova, G.M.Kaleva, M.V.Kudinova, S.G.Prutchenko et al., The influence of non-stoichiometry on crystal structure and morphology of high-temperature superconductive $YBa_2Cu_3O_7$ ceramic, *Ferroelectrics*, **105** [1/4] 45-50 (1990).

[14]K.I.Furaleva, G.M.Kaleva, S.G.Prutchenko, V.V.Kolesov, E.D.Politova, Microstructure and critical currents of the modified $YBa_2Cu_3O_{7-\delta}$ ceramics, *Russian Journal of Inorganic Materials,* **33** [10], 1276-1280 (1997).

[15]E.D.Politova, High-temperature superconducting ceramic on the basis of the yttrium-barium and bismuth-strontium-calcium cuprates, *Russian Chemical Journal*, **XLII** [4] 91-100 (1998).

[16]S.R.Ovshinsky, R.T.Young, D.D.Allred, G.DeMaggio, G.A.Van der Leeden, Superconductivity at 155 K, *Phys.Rev.Letters*, **58** [24] 2579-81 (1987).

[17]V.L.Ginzburg, On two dimensional superconductors, *Physica Scripta*, **27** [1] 76-77 (1989).

[18]G.M.Kaleva, E.D.Politova, M.V.Kudinova, S.G.Prutchenko, Yu.N.Venevtsev, Superconducting properties of the modified $YBa_2Cu_3O_7$ ceramic, *Russian Journal of Inorganic Materials*, **28 [8]** 1776-80 (1992).

[19]K.I.Furaleva, S.G.Prutchenko, E.D.Politova, Microstructure and superconducting properties of the doped $YBa_2Cu_3O_7$ ceramics, *Russian Journal on Superconductivity: Physics, Chemistry, Technique*, **8** [5-6], 702-8 (1996).

[20]D.E.Zherebchevskii, T.N.Moiseeva, B.Ya.Sukharevskii et al., The form of the surface impedance line at the superconducting transition in metlloceramics on base of yttrium. *Russian Journal of Low Temperature Physics,* **15** [1] 106-10 (1989); Absorption of electromagnet waves in the HTSC metallooxides on base of yttrium and lanthanoids, *Ibid,* **15** [7] 695-709 (1989).

[21]J.R.Clem, Two dimensional vortices in a stack of thin superconducting films: a model for high temperature superconducting multilayers, *Phys. Rev. B*, **43** [10] 7837-46 (1991); Fundamentals of vortices in the high temperature superconductors, *Supercond. Sci. Technol.* 5 [1] 33-40 (1991); Anisotropy and two-dimensional behaviour in the high-temperature superconductors, *Ibid* 11 [10] 909-14 (1998); Phenomenological theory of magnetic structure in the high temperature superconductors, *Physica C* **162-164** 1137-42 (1989).

[22]V.N.Varyukhin, V.V.Chabanenko, High frequency properties of the superconducting metaloxide, *Russian Journal of Superconductivity: Physics, Chemistry, Technique*, **3** [1] 73-78 (1990); High frequency losses and intergrain coupling in high –Tc superconductors, *Ibid,* **4** [9] 1821-34 (1991).

THE CRYSTAL STRUCTURES OF SOME TRANSITION METAL STABILISED MERCURY CUPRATE SUPERCONDUCTORS

Neil C. Hyatt*
Department of Engineering Materials,
The University of Sheffield,
Mappin Street,
Sheffield, S1 3JD. UK.

Ian Gameson
School of Chemistry,
The University of Birmingham,
Edgbaston,
Birmingham, B15 2TT. UK.

ABSTRACT

The synthesis, crystal structures and superconducting properties of three families of transition metal stabilised mercury cuprates are discussed. The crystal structures of $(Hg_{0.60}V_{0.40})Sr_2(Nd_{0.53}Sr_{0.47})Cu_2O_{6.94}$, $(Hg_{0.78}Re_{0.22})Sr_2Nd_{(0.59}Ca_{0.41})Cu_2O_{6.88}$ and $(Hg_{0.75}Mo_{0.25})Sr_2CuO_{4.6}$ were determined from Rietveld analysis of X-ray and neutron powder diffraction data. These data demonstrate that the linear HgO_2^{2-} units in $(Hg_{0.60}V_{0.40})Sr_2(Nd_{0.53}Sr_{0.47})Cu_2O_{6.94}$, $(Hg_{0.78}Re_{0.22})Sr_2Nd_{(0.59}Ca_{0.41})Cu_2O_{6.88}$ and $(Hg_{0.75}Mo_{0.25})Sr_2CuO_{4.6}$, are partially substituted by tetrahedral VO_4^{3-}, octahedral ReO_6^{6-} and tetrahedral MoO_4^{2-} species, respectively. Hole doping in these systems arises from a combination of oxyanion substitution, cation substitution and intercalation of labile oxygen anions.

INTRODUCTION

The synthesis of members of the $HgSr_2Ca_{n-1}Cu_nO_{2n+2+\delta}$ system has not yet proved possible, even under the forcing conditions of high-pressure methods [1, 2]. However, the partial substitution of mercury by a high-valent transition metal (M) allows the solid state synthesis of substituted derivatives, formulated $(Hg_{1-x}M_x)Sr_2(Ln_{1-y}Ca_y)Cu_2O_{6+\delta}$ (Ln = trivalent lanthanide), at room pressure [2 – 4]. It is believed that the structure stabilising influence of the transition metal element, in this context, arises from the additional oxygen atoms associated with the local octahedral / tetrahedral co-ordination environment of the transition metal cation [4]. Although a wide range of $(Hg_{1-x}M_x)Sr_2(Ln_{1-y}Ca_y)Cu_2O_{6+\delta}$ derivatives have been reported (with M = Ti, V, Cr, Nb, Mo, Ta, W and Re) the detailed structural characterisation of these compounds has proved problematic, since near single-phase specimens of these materials are rather difficult to prepare. However, such structural investigations are essential to fully understand the role of the transition metal cation in stabilising these materials and to clarify the effect of different transition metal cations on the superconducting properties of these chemically complex compounds.

EXPERIMENTAL

The synthesis of polycrystalline materials with the nominal compositions $(Hg_{0.7}V_{0.3})Sr_2(Nd_{1-y}Sr_y)Cu_2O_{6+\delta}$ (0.3 < y < 0.5) and $(Hg_{0.8}Re_{0.2})Sr_2Nd_{(1-y}Ca_y)Cu_2O_{6+\delta}$ (0.3 < y < 0.5) was achieved by solid state reaction between stoichiometric quantities of HgO, V_2O_5, ReO_3, MoO_3, Nd_2O_3, $SrCuO_2$ or Sr_2CuO_3, and CaO or SrO_2, as appropriate. These reagents were intimately mixed and pressed into pellets which were wrapped in gold foil and sealed in a quartz ampoule. The quartz ampoule was typically heated to 880°C for 16 h and subsequently quenched to room temperature. The $SrCuO_2$ and Sr_2CuO_3 precursors were obtained by calcining stoichiometric quantities of $SrCO_3$ and CuO at 950°C in air, for 48 h.

X-ray powder diffraction data were acquired at room temperature using a Siemens D5000 diffractometer, equipped with a primary beam curved single-crystal Ge-(220) monochromator, affording Cu-Kα_1 radiation. The instrument operates in transmission mode and is fitted with a Position Sensitive Detector (PSD). During data collection, specimens (consisting of a thin layer of sample dispersed on Mylar film) were rotated in order to alleviate preferred orientation effects. For the purpose of structure refinement studies, the diffraction data were corrected for sample absorption effects using the method described by Klug and Alexander [5]. Time-of-flight neutron powder diffraction studies were performed using the POLARIS diffractometer at the ISIS facility, Rutherford Appleton Laboratory, UK. For the purpose of Rietveld profile analysis, data from the high-resolution C-bank detectors ($135° < 2\theta < 160°$, $\Delta d / d = 0.5 \%$) were analysed (the data were corrected for sample attenuation). Structure refinement was undertaken using the GSAS suite of programs [6], using the following general strategy. In the initial step, a suitable model was constructed based on the archetype $HgBa_2CaCu_2O_{6+\delta}$ [7] or $HgBa_2CuO_{4+\delta}$ [8] structure, modified according to the nominal stoichiometry, no interstitial oxygen sites were assumed in the $(Hg,M)O_\delta$ plane. After obtaining an adequate fit to the data on the basis of the initial model, a difference Fourier analysis was undertaken in order to identify oxygen atoms co-ordinated to the transition metal cations in the $(Hg,M)O_\delta$ plane. The model was subsequently modified to accommodate these oxygen atoms and provide for an appropriate tetrahedral or octahedral co-ordination environment about the transition metal cation. The co-ordinates, isotropic thermal parameters and fractional occupancies of all atoms were then refined, as appropriate, subject to suitable constraints: mixed site occupancies were constrained to sum to unity; the occupancies of oxygen sites co-ordinated to transition metal cations were constrained to maintain an appropriate co-ordination number (four or six); and the isotropic thermal parameters of atoms with constrained site occupancies were constrained to be equal. In the final step, the possibility of an interstitial oxygen site, with co-ordinates (0.5, 0.5, 0), was tested. Bond Valence Sums for selected cations were calculated using the scheme of Brown and Altermatt [9].

RESULTS AND DISCUSSION

The $(Hg_{0.7}V_{0.3})Sr_2(Nd_{1-y}Sr_y)Cu_2O_{6+\delta}$ system

X-ray powder diffraction indicated the formation of $(Hg_{0.65}V_{0.35})Sr_2(Nd_{1-y}Sr_y)Cu_2O_{6+\delta}$ materials in the narrow nominal composition range $0.3 \leq y \leq 0.5$. Compositions with $y = 0.4$ and $y = 0.5$ were found to be superconducting with $T_c = 67$ K and $T_c = 82$ K, respectively; in contrast, samples with $y = 0.3$ were found to be non-superconducting. A near single-phase sample of nominal composition $(Hg_{0.65}V_{0.35})Sr_2(Nd_{0.6}Sr_{0.4})Cu_2O_{6+\delta}$ was selected for detailed structural analysis. In order to determine accurate structural parameters for this material, it was necessary to undertake a combined X-ray and neutron powder diffraction study, due to the small X-ray scattering factor of oxygen and the very small neutron scattering length of vanadium ($b_c = -0.3824$ barns). The co-ordination environment of the vanadium cation was determined directly from a ^{51}V solid state NMR study. The ^{51}V Magic Angle Spinning (MAS) and static NMR spectra of the sample of nominal composition $(Hg_{0.65}V_{0.35})Sr_2(Nd_{0.6}Sr_{0.4})Cu_2O_{6+\delta}$ showed an isotropic line-shape centred at $\delta_{iso} = -507$ ppm. The MAS and static line-widths of 2 KHz and 12 KHz, respectively (given as full width at half-height maximum), together with the isotropic chemical shift, are characteristic of an undistorted tetrahedral vanadium co-ordination environment [10], indicating the presence of isolated VO_4^{3-} oxyanions in the crystal structure of this material. The refined composition of the sample was determined to be $(Hg_{0.60}V_{0.40})Sr_2(Nd_{0.53}Sr_{0.47})Cu_2O_{6.94}$, close to the nominal composition. The refined structural parameters of $(Hg_{0.60}V_{0.40})Sr_2(Nd_{0.53}Sr_{0.47})Cu_2O_{6.94}$ are given in Table I and the final profile fit to the X-ray and neutron diffraction data are shown in Figures 1a and 1b, respectively.

The crystal structure of $(Hg_{0.60}V_{0.40})Sr_2(Nd_{0.53}Sr_{0.47})Cu_2O_{6.94}$, Figure 1c, may be considered as being composed of a regular intergrowth of rock-salt type $(HgO_\delta)(SrO)(HgO_\delta)$ and (oxygen deficient) perovskite type $(CuO_2)(Nd)(CuO_2)$ layers, with partial substitution of Nd^{3+} by Sr^{2+} and HgO_2^{2-} units by tetrahedral VO_4^{3-} oxyanions. The limited Nd / Sr solid solution range in the $(Hg_{0.65}V_{0.35})Sr_2(Nd_{1-y}Sr_y)Cu_2O_{6+\delta}$ system ($0.3 \leq y \leq 0.5$) is considered to arise from the significant size mismatch of the Nd / Sr cations, present in the 8-fold co-ordinate site of the perovskite type layer. The VO_4^{3-} oxyanions in $(Hg_{0.60}V_{0.40})Sr_2(Nd_{0.53}Sr_{0.47})Cu_2O_{6.94}$ are of near ideal tetrahedral symmetry, within the errors associated with the refined vanadium-oxygen bond lengths and bond angles. The average V-O bond length, 1.73 Å, is close to that found in other isolated VO_4^{2-} tetrahedra; in Li_3VO_4, for example, the average V-O bond length is 1.72 Å [11]. The substitution of linear HgO_2^{2-} units by VO_4^{3-} tetrahedra in $(Hg_{0.60}V_{0.40})Sr_2(Nd_{0.53}Sr_{0.47})Cu_2O_{6.94}$, results in a significant distortion of the neighbouring CuO_5 and $SrO_{8+\delta}$ polyhedra. The Bond Valence Sum of the vanadium cation, 4.9+, indicates that the bonding environment of this cation is essentially optimal.

Annealing the sample with the refined composition $(Hg_{0.60}V_{0.40})Sr_2(Nd_{0.53}Sr_{0.47})Cu_2O_{6.94}$ under flowing oxygen, at 300 °C for 24 h, did not change T_c significantly from the value determined for the as-prepared material, $T_c = 67$ K. In contrast, annealing this sample under flowing argon, at 300 °C for 24 h, reduced the T_c of the as-prepared material to 54 K. The refined composition of the as-prepared material indicates that ~94 % of the available interstitial oxygen sites are occupied. Approximately 80 % of the available interstitial sites are occupied by non-labile oxygen atoms co-ordinated to vanadium cations, in the form of VO_4^{3-} tetrahedra. Thus, annealing the as-prepared material under oxygen does not lead to any further intercalation of additional oxygen atoms, since there are few available interstitial sites to accommodate such labile species. However, it would appear that annealing under argon is effective in removing a small fraction of the labile oxygen atoms present in the as-prepared material, leading to a more under-doped material with a reduced transition temperature. The maximum $T_c = 82$ K observed in the $(Hg_{0.65}V_{0.35})Sr_2(Nd_{1-y}Sr_y)Cu_2O_{6+\delta}$ system for the nominal composition y = 0.5, is close to that reported for similar nominal compositions in the $(Hg_{0.65}V_{0.35})Sr_2(Nd_{1-y}Ca_y)Cu_2O_{6+\delta}$ system, with $T_c = 80$ K for y = 0.55 [12], and the $(Hg_{0.7}V_{0.3})Sr_2(Y_{1-y}Ca_y)Cu_2O_{6+\delta}$ system, with $T_c = 80$ K for y = 0.5 [13] . The similar size of the Nd^{3+} and Ca^{2+} cations (ionic radii are 1.11 Å and 1.12 Å, respectively, for 8-fold co-ordination [14]) affords a wide composition range in the $(Hg_{0.65}V_{0.35})Sr_2(Nd_{1-y}Ca_y)Cu_2O_{6+\delta}$, system with $0.20 \leq y \leq 0.80$. This composition range allows T_c to be varied smoothly in this system, through control of the Nd / Ca ratio, with T_c reaching a maximum of 105 K for y = 0.65. In contrast, the significant size mismatch between the Nd^{3+} and Sr^{2+} cations affords a smaller composition range in the $(Hg_{0.65}V_{0.35})Sr_2(Nd_{1-y}Sr_y)Cu_2O_{6+\delta}$ system, with $0.3 \leq y \leq 0.5$. This composition range effectively inhibits the chemical control of superconductivity in this system, through variation of the Nd / Sr ratio, since only nominally underdoped compositions may be prepared.

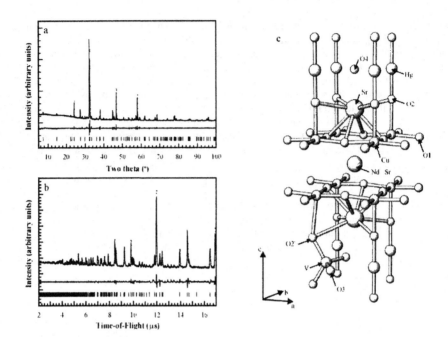

Figure 1: Final profile fit (solid line) to a) X-ray and b) neutron powder diffraction data (solid points) for $(Hg_{0.60}V_{0.40})Sr_2(Nd_{0.53}Sr_{0.47})Cu_2O_{6.94}$, the difference profile is shown below and the tick marks show the positions of the allowed Bragg reflections; c) schematic illustration of the crystal structure of $(Hg_{0.60}V_{0.40})Sr_2(Nd_{0.53}Sr_{0.47})Cu_2O_{6.94}$.

The $(Hg_{0.8}Re_{0.2})Sr_2(Nd_{1-y}Ca_y)Cu_2O_{6+\delta}$ system

A systematic study of the $(Hg_{0.8}Re_{0.2})Sr_2(Nd_{1-y}Ca_y)Cu_2O_{6+\delta}$ solid solution revealed that phase purity was essentially independent of y in the composition range $0.2 \leq y \leq 0.6$. Weak reflections not consistent with space group P4/mmm were observed in the X-ray powder diffraction patterns of all $(Hg_{0.8}Re_{0.2})Sr_2(Nd_{1-y}Ca_y)Cu_2O_{6+\delta}$ compositions in this range. A careful analysis of these reflections provided no evidence for the existence of a superstructure arising from Hg / Re ordering, as has been reported for $(Hg_{1-x}Re_x)Sr_2Ca_{n-1}Cu_nO_{2n+2+\delta}$ materials (n = 2, 3) prepared *via* high pressure methods [15]. These additional reflections were thus attributed to unidentified impurity phases. Vacuum annealed compositions with $0.3 \leq y \leq 0.5$ were found to be superconducting with a maximum $T_c = 100$ K, see Figure 2.

For the purpose of structure refinement, neutron powder diffraction data were acquired on a sample with nominal composition $(Hg_{0.8}Re_{0.2})Sr_2(Nd_{0.6}Ca_{0.4})Cu_2O_{6+\delta}$. The refined composition, $(Hg_{0.78}Re_{0.22})Sr_2(Nd_{0.59}Ca_{0.41})Cu_2O_{6.88}$ was found to be essentially identical to the nominal composition. The final profile fit to the neutron powder diffraction data is shown in Figure 2a and the refined structural parameters for $(Hg_{0.78}Re_{0.22})Sr_2(Nd_{0.59}Ca_{0.41})Cu_2O_{6.88}$ are reported in Table II.

Table I : Refined structural parameters for $(Hg_{0.60}V_{0.40})Sr_2(Nd_{0.53}Sr_{0.47})Cu_2O_{6.94}$.

Space group:	P4/mmm				
Lattice parameters	$a = 3.8765(1)$ Å		$c = 11.9232(5)$ Å		
Atom	Site	x	y	z	n	B_{iso} (Å^2)
Hg	1 a	0	0	0	0.600(15)	1.3(1)
V	4 l	0.16(1)	0	0	0.100(4)	1.3(1)
Sr	2 h	0.5	0.5	0.2075(2)	1	0.72(4)
Nd	1 d	0.5	0.5	0.5	0.53(2)	0.27(5)
Sr	1 d	0.5	0.5	0.5	0.47(2)	0.27(5)
Cu	2 g	0	0	0.3544(2)	1	0.36(3)
O1	4 i	0.5	0.0	0.3633(2)	1	0.51(4)
O2	2 g	0	0	0.1643(9)	0.600(15)	1.8(8)
O2'	8 s	0.0324(16)	0	0.126(3)	0.100(4)	1.8(8)
O3	4 j	0.379(2)	0.379(2)	0	0.200(2)	1.5(2)
O4	1 c	0.5	0.5	0	0.14(1)	1.6(2)

Powder statistics:	R_{wp} (%)	R_p (%)	
X-ray data	6.43	4.42	
Neutron data	1.96	4.27	
Combined data	4.76	4.41	$\chi^2 = 8.61$

Constraints:	Atom fractions:	n (Hg) + 4n (V) = 1, n (O2) + 4n (O2') = 1, n (O2') = n (V),
		2n (O3) = (V), n (Nd) + n (Sr) = 1
	Thermal parameters:	B_{iso}(Hg) = B_{iso}(V), B_{iso} (O2) = B_{iso} (O2'), B_{iso}(Nd) = B_{iso}(Sr)

Table II: Refined structural parameters for $(Hg_{0.88}Re_{0.22})Sr_2(Nd_{0.59}Ca_{0.41})Cu_2O_{6.88}$.

Space group:	P4/mmm				
Unit-cell parameters:	$a = 3.84028(4)$ Å		$c = 11.9322(2)$ Å		
Atom	Site	x	y	z	n	B_{iso} (Å^2)
Hg	1 a	0	0	0	0.78(1)	0.70(1)
Re	1 a	0	0	0	0.22(1)	0.70(1)
Sr	2 h	0.5	0.5	0.2087(1)	1	0.44(2)
Nd	1 d	0.5	0.5	0.5	0.59(1)	0.17(2)
Ca	1 d	0.5	0.5	0.5	0.41(1)	0.17(2)
Cu	2 g	0	0	0.35825(9)	1	0.23(2)
O1	4 i	0.5	0	0.36807(8)	1	0.34(1)
O2	2 g	0	0	0.1631(1)	1	1.13(3)
O3	4 j	0.3482(9)	0.3482(9)	0	0.22(1)	0.38(4)

Powder statistics:	R_{wp} = 3.30 %	R_p = 6.34 %	$\chi^2 = 5.15$ %
Constraints:	n(Re) + n(Hg) = 1; n(Nd) + n(Ca) = 1; n(O3) = 4n(Re); B_{iso}(Nd) = B_{iso}(Ca)		

Table III: Refined structural parameters for $(Hg_{0.75}Mo_{0.25})Sr_2CuO_{4.6}$.

Space group:	P4/mmm				
Lattice parameters	$a = 3.7794(1)$ Å		$c = 8.8600(3)$ Å		
Atom	Site	x	y	z	n	B_{iso} (Å^2)
Hg	1 a	0	0	0	0.748(3)	0.7(1)
Mo	4 l	0.11(1)	0	0	0.063(1)	0.7(1)
Sr	2 h	0.5	0.5	0.2958(2)	1	0.3(1)
Cu	2 g	0	0	0.5	1	0.3(1)
O1	4 i	0.5	0.0	0.5	1	0.5(1)
O2	2 g	0	0	0.2225(3)	0.748(3)	0.5(2)
O2'	8 s	0.134(2)	0	0.169(1)	0.063(1)	0.5(2)
O3	4 j	0.376(2)	0.376(2)	0	0.126(1)	0.6(2)
O4	1 c	0.5	0.5	0	0.10(1)	0.6(2)

Powder statistics:	R_{wp} = 3.55%	R_p = 2.90 %	$\chi^2 = 4.35$ %
Constraints:	Atom fractions:	n (Hg) + 4n (Mo) = 1, n (O2) + 4n (O2') = 1	
		n (O2') = n (Mo), 2n (O3) = n (Mo)	
	Thermal parameters:	B_{iso}(Hg) = B_{iso}(Mo), B_{iso} (O2) = B_{iso} (O2')	

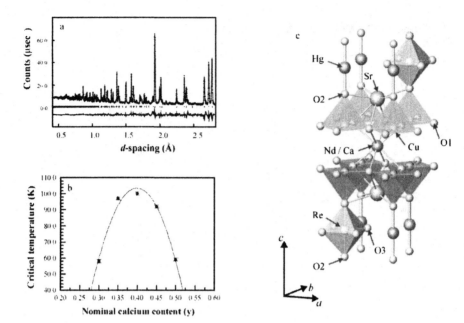

Figure 2: a) Final profile fit (solid line) to the neutron powder diffraction data of $(Hg_{0.88}Re_{0.22})Sr_2(Nd_{0.59}Ca_{0.41})Cu_2O_{6.88}$ (solid points); the tick marks denote the positions of the allowed Bragg reflections, the difference profile is shown by the lower solid line. b) Variation of T_c with Ca content in the $(Hg_{0.8}Re_{0.2})Sr_2(Nd_{1-y}Ca_y)Cu_2O_{6+\delta}$ system. c) Schematic illustration of the crystal structure of $(Hg_{0.88}Re_{0.22})Sr_2(Nd_{0.59}Ca_{0.41})Cu_2O_{6.88}$.

The crystal structure of $(Hg_{0.78}Re_{0.22})Sr_2(Nd_{0.59}Ca_{0.41})Cu_2O_{6.88}$, Figure 2c, may be considered as being composed of a regular intergrowth of rock-salt type $(HgO_\delta)(SrO)(HgO_\delta)$ and (oxygen deficient) perovskite type $(CuO_2)(Nd)(CuO_2)$ layers, with partial substitution of Nd^{3+} by Ca^{2+} and $HgO_2{}^{2-}$ units by octahedral $ReO_6{}^{6-}$ species. The in-plane Re-O3 bond length of the ReO_6 octahedron, 1.891(5) Å, is close to that found in the related phase $Hg_{0.75}Re_{0.25}Sr_2CaCu_2O_7$, 1.883(11) Å [16], and in agreement with the sum of the ionic radii of these elements (0.55 Å for Re^{6+}, 1.35 Å for O^{2-}, [14]).

Hole doping of the CuO_2 sheets in vacuum annealed samples of the $(Hg_{0.8}Re_{0.2})Sr_2(Nd_{1-y}Ca_y)Cu_2O_{6+\delta}$ system ($0.3 \le y \le 0.5$) may be sensitively controlled by suitable adjustment of the Nd / Ca ratio. This is demonstrated by Figure 2b, in which the variation of the superconducting transition temperature with the nominal calcium content is observed to track the familiar 'bell-shaped' curve observed in a wide range of cuprate superconductors [17]. With increasing calcium content the transition from the under-doped (y < 0.4) to over-doped (y > 0.4) regime in $(Hg_{0.8}Re_{0.2})Sr_2(Nd_{1-y}Ca_y)Cu_2O_{6+\delta}$ may be controlled, with optimal doping and maximum $T_c \sim 100$ K observed at $y \sim 0.4$. Assuming the oxidation state of rhenium is Re^{6+}, the average copper oxidation state for the refined composition $(Hg_{0.78}Re_{0.22})Sr_2(Nd_{0.59}Ca_{0.41})Cu_2O_{6.88}$ is 2.15+. This is close to the copper valence of 2.16 - 2.18+ corresponding to optimal doping in a wide range of cuprates [17] and is consistent with the data presented in Figure 2c which indicate that the highest T_c observed in the $(Hg_{0.8}Re_{0.2})Sr_2(Nd_{1-y}Ca_y)Cu_2O_{6+\delta}$ system occurs near the nominal composition with $y = 0.4$.

$(Hg_{0.75}Mo_{0.25})Sr_2CuO_{4.6}$

The crystal structure of $(Hg_{0.75}Mo_{0.25})Sr_2CuO_{4.6}$ was determined by Rietveld analysis of neutron powder diffraction data. The final profile fit is shown in Figure 3a and the refined structural parameters for $(Hg_{0.75}Mo_{0.25})Sr_2CuO_{4+\delta}$ are reported in Table III. The crystal structure of $(Hg_{0.75}Mo_{0.25})Sr_2CuO_{4.6}$ is similar to that of the barium analogue $(Hg_{0.75}Mo_{0.25})Ba_2CuO_{4+\delta}$ [18] and the archetype $HgBa_2CuO_{4+\delta}$ [8], with complete replacement of Ba by Sr and partial substitution of linear HgO_2^{2-} units by isolated tetrahedral MoO_4^{2-} oxanions, as shown in Figure 3b. The MoO_4^{2-} oxanions in $(Hg_{0.75}Mo_{0.25})Sr_2CuO_{4.6}$ are of near ideal tetrahedral symmetry, within the errors associated with the refined molybdenum-oxygen bond lengths and bond angles.

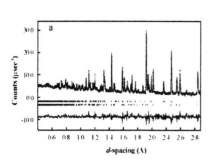
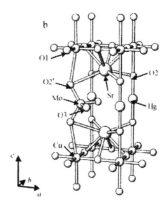

Figure 3: a) Final profile fit (solid line) to the neutron powder diffraction data of $(Hg_{0.75}Mo_{0.25})Sr_2CuO_{4.6}$ (solid points) (right); the tick marks denote the positions of the allowed Bragg reflections, the difference profile is shown by the lower solid line. b) Schematic illustration of the crystal structure of $(Hg_{0.75}Mo_{0.25})Sr_2CuO_{4.6}$.

The average Mo-O bond length in $(Hg_{0.75}Mo_{0.25})Sr_2CuO_{4.6}$, 1.75 Å, is close to that found in other isolated MoO_4^{2-} tetrahedra; for example, the five different MoO_4 tetrahedra in $La_2(MoO_4)_3$ and $Ce_6(MoO_4)_8(Mo_2O_7)$ have average Mo-O bond lengths in the range 1.77 – 1.78 Å and 1.75 – 1.77 Å, respectively [19]. The bond valence sum of the Mo cation, 6.1+, indicates that the tetrahedral bonding environment of this cation is essentially optimal. The substitution of linear HgO_2^{2-} units by MoO_4^{2-} tetrahedra in $(Hg_{0.75}Mo_{0.25})Sr_2CuO_{4.6}$ results in a significant distortion of the neighbouring CuO_6 and $SrO_{8+\delta}$ polyhedra. Since the HgO_2^{2-} unit and MoO_4^{2-} oxanions in $(Hg_{0.75}Mo_{0.25})Sr_2CuO_{4.6}$ are isovalent, the partial replacement of Hg by Mo is not effective in doping the CuO_2 sheets in the $(Hg_{1-x}Mo_x)Sr_2CuO_{4+\delta}$ system. The mechanism of hole doping in the $(Hg_{1-x}Mo_x)Sr_2CuO_{4+\delta}$ system is therefore through intercalation of interstitial oxygen atoms, as in the archetype $HgBa_2CuO_{4+\delta}$.

REFERENCES

1. B. Raveau, C. Michel, M. Hervieu and A. Maignan, "Crystal chemistry of superconducting mercury-based cuprates and oxycarbonates", *J. Mater. Chem.*, **5** 803 (1995).

2. S. Hahakura, J. Shimoyama, O. Shiino and K. Kishio, "New barium-free mercury based high T_c superconductors $(Hg,Mo)Sr(Ca,Y)_{n-1}Cu_nO_y$ and $HgSr_2(Ca,Y)_{n-1}(Cu,Re)_nO_y$ (n = 1 and 2), *Physica C*, **233** 1 (1994).

3. S. Hahakura, J. Shimoyama, O. Shiino and K. Kishio, "Chemical stabilisation of Hg-based superconductors", *Physica C*, **235** 915 (1994).

4. N.C. Hyatt, G.B. Peacock, I. Gameson, K.L. Moran, M. Slaski, M.O. Jones, A.J. Ellis, Y.E. Gold, R. Dupree and P.P. Edwards, "On the role of transition metal elements as structure-stabilising agents in cuprate superconductors", *Int. J. Inorg. Mater.*, **1** 87 (1999).

5. H.P. Klug and L.E. Alexander, "X-ray Diffraction Procedures for Polycrystalline and Amorphous Materials", John Wiley and Sons, (1996).

6. A.C. Larson and R.B. Von Dreele, "GSAS – General Structural Analysis System", Report LA-UR-86-748, Los Alamos National Laboratory, Los Alamos, NM 87545, (1990).

7. P.G. Radaelli, J. Wagner, B.A. Hunter, M.A. Beno, G.S. Knapp, J.D. Jorgensen, D.G. Hinks, "Structure, doping and superconductivity in $HgBa_2CaCu_2O_{6+\delta}$ ($T_c \leq 128$ K)", *Physica C*, **216** 29 (1993).

8. S.N. Putlin, E.V. Antipov, O. Chmaissem and M. MArezio, "Superconductivity at 94 K in $HgBa_2CuO_{4+\delta}$", *Nature*, **362** 226 (1993).

9. I.D. Brown and S. Altermatt, "Bond-valence parameters obtained from a systematic analysis of the Inorganic Crystal Structure Database", *Acta Cryst.*, **B41** 244 (1985).

10. O.B. Lapina, V.M. Mastikhin, A.A. Shubin, V.N. Krasilnikov and K.I. Zamaraev," V-51 solid state NMR studies of vanadia based catalysts", *Prog. In NMR Spectro.*, **24** 457 (1992).

11. R.D. Shannon and C. Calvo, "Refinement of the crystal structure of low temperature Li_3VO_4 and analysis of mean bond lengths in phosphates, arsenates and vanadates", *J. Solid State Chem.*, **6** 538 (1973).

12. E. Kandyel, T. Kamiyama, H. Asano, S. Amer and M. Abou-Sekkina, "Synthesis, structure and superconductivity of $(Hg_{0.65}V_{0.35})Sr_2(Ca_{1-x}Nd_x)Cu_2O_z$", *Jpn. J. Appl. Phys.*, **36** 6306 (1997).

13. V. Badri, Y.T. Wang, J. Onstad and A.M. Hermann, "Synthesis and physical properties of the Hg-based superconductors with the 1212 structure: $Hg_{0.7}V_{0.3}Sr_2Ca_{1-x}Y_xCu_2O_{6+\delta}$" *J. Supercon.*, **10** 563 (1997).

14. R.D. Shannon, "Revised effective ionic radii and systematic studies of interatomic distances in halides and chalcogenides", *Acta. Cryst.*, **A32** 751 (1976).

15. K. Yamaura, J. Shimoyama, S. Hahakura, Z. Hiroi, M. Takano and K. Kishio, "High-pressure synthesis and superconductivity of a Ba free mercury-based superconductor $(Hg_{0.75}Re_{0.25})Sr_2Ca_2Cu_3O_y$", *Physica C*, **246** 351 (1995).

16. O. Chmaissem, J.D. Jorgensen, K. Yamaura, Z. Hiroi, M. Takano, J. Shimoyama and K. Kishio, "Crystal structures of Hg-Sr-Ca-Cu-O superconductors with enhanced flux pinning: $Hg_{1-x}Re_xSr_2Ca_{n-1}Cu_nO_{2n+2+\delta}$ (n = 2, 3; x ≈ 0.2 - 0.25)", *Phys. Rev. B*, **53** 14647 (1996).

17. J.L. Tallon, C. Bernhard, H. Shaked, R.L. Hitterman and J.D. Jorgensen, "Generic superconducting phase behaviour in High T_c cuprates – T_c variation with hole concentration in $YBa_2Cu_3O_{7-\delta}$", *Phys. Rev. B*, **51** 12911 (1995).

18. F. Licci, M. Marezio, Q. Huang, A. Santoro, C. Bougerol-Challiot and R. Masini, "Synthesis, structure and superconductivity of $Hg_{0.75}Mo_{0.25}Ba_2CuO_{4+\delta}$", *Physica C*, **325** 41 (1999).

19. B.M. Gatehouse and R. Same, "The crystal structure of a complex cerium (III) molybdate; $Ce_6(MoO_4)_8(Mo_2O_7)$", *J. Solid State Chem.*, **25** 115 (1978).

Author Index

Keyword Index